Cryptanalytic Attacks on RSA

Cryptanalytic Attacks on RSA

by

Song Y. Yan
University of Bedfordshire, UK
and
Massachusetts Institute of Technology, USA

Song Y. Yan, PhD
Professor of Computer Science and Mathematics
Director, Institute for Research in Applicable
Computing
University of Bedfordshire
Bedfordshire LU1 3JU
UK
song.yan@beds.ac.uk
and
Visiting Professor
Department of Mathematics
Massachusetts Institute of Technology
Cambridge, MA 02139-4307
USA
syan@math.mit.edu

ISBN-13: 978-1-4419-4310-1 e-ISBN-13: 978-0-387-48742-7

Printed on acid-free paper.

9 8 7 6 5 4 3 2 1

springer.com

Dedicated to Professor Glyn James
on the occasion of his 70th birthday
with gratitude for his encouragement and support

Table of Contents

Preface

The art of war teaches us to rely not on the likelihood of the enemy's not coming, but on our own readiness to receive him; not on the chance of his not attacking, but rather on the fact that we have made our position unassailable.

SUN TZU
The Art of War (500 BC)

The book is about the cryptanalytic attacks on RSA. RSA is the first *workable* and *practical* public-key cryptographic system, invented in 1977 and published in 1978, by Rivest, Shamir and Adleman, then all at the Massachusetts Institute of Technology (MIT), and is still the most widely used cryptographic systems in e.g., online transactions, emails, smartcards, and more generally electronic and mobile commerce over the Internet, for which its three inventors received the year 2002 Turing Award, a prize considered to be the equivalent *Nobel Prize for Computer Science.* The security of RSA relies on the computational intractability of the Integer Factorization Problem (IFP), for which, no efficient (i.e., polynomial-time) algorithm is known. To get an idea how difficult the *integer factorization* is, let us consider the following 2048 bits (617 digits) composite number, known as RSA-2048:

251959084756578934940271832400483985714292821262040320277771378360_
436620207075955562640185258807844069182906412495150821892985591491_
761845028084891200728449926873928072877767359714183472702618963750_
149718246911650776133798590957000973304597488084284017974291006424_
586918171951187461215151726546322822168699875491824224336372590851_
418654620435767984233871847744479207399342365848238242811981638150_
106748104516603773060562016196762561338441436038339044149526344321_
901146575444541784240209246165157233507787077498171257724679629263_
863563732899121548314381678998850404453640235273819513786365643912_
12010397122822120720357.

It is a *product* of two *prime numbers.* The RSA Data Security Incorporation currently offers a $200,000 prize for the first person or group finding its two prime factors. The basic idea of RSA encryption and decryption is, surprisingly, rather simple:

$$C \equiv M^e \pmod{N}, \quad M \equiv C^d \pmod{N},$$

where $N = pq$ with p and q prime, M, C, e and d are the plaintext, ciphertext, encryption exponent and decryption exponent, respectively. Note that e and d must be satisfied with the condition that $ed \equiv 1 \; (\bmod \; \phi(N))$, where $\phi(N) = (p-1)(q-1)$ is Euler's ϕ-function. Let, for example, $e = 65537$, N be the above mentioned number RSA-2048, and C the following number:

218598056144555493024019389629177159753811144728543422921500499254_
181211032562087679022259831067991286101190897695119357754765408522_
697956638242922870637083231694404873947694078432775781998614979942_
064361669462614088852741600217233052059574880668463536030287944235_
822627708134997061064700771693064600712629809165416998449992925313_
374281387325903328781863209595468701560742767599157207314869432305_
892651836189508103764678721683360183118994273706398707795480800698_
501878875875150532123738006235671958527639461339868604410378449818_
383913059864587128396200112815989134558427750667427151537609736712_
04647757116059031684587.

To recover M from C one requires to find d; to find d one needs to calculate $\phi(N)$; to calculate $\phi(N)$ one needs to factor N. But unfortunately, factorizing N is intractable when N is large (in the present case, N is a 2048-bit number, which is far beyond the computing power of any factoring algorithm on any computer at present); no polynomial-time factoring algorithm is known so far. Thus, RSA is secure and C is safe since it is difficult to recover M from C without factoring N. This is essentially the whole idea of RSA! One can try to decrypt the above given RSA ciphertext C or try to factor the number RSA-2048 in order to get an idea how difficult it is to break RSA or to factor a large number.

The book consists of ten chapters. Chapter 1 presents some computational and mathematical preliminaries, particularly the theory and practice of tractable and intractable computations in number theory. Chapter 2 introduces the basic concepts and theory of the RSA cryptographic system and its variants in a broad sense. As the security of RSA is based on the intractability of the Integer Factorization Problem (IFP), which is also closely related to the Discrete Logarithm Problem (DLP), the attacks based on solutions to IFP problem are discussed in Chapter 3, whereas the attacks based on solutions to DLP problem are discussed in Chapter 4. As quantum algorithm is applicable to both the IFP problem and the DLP problem, Chapter 5 will discuss some quantum attacks on RSA via quantum order finding, quantum factoring and quantum discrete logarithm solving. Chapter 6 concentrates on some simple elementary number-theoretic attacks on RSA, including e.g., forward attack, short plaintext attack, common modulus attack and fixed-point attack. It is common that to speed-up the computation of RSA encryption, a short public exponent e is often used. It is also true for the RSA decryption if a short private exponent d is used. However, the use of short exponent e or d can be dangerous. So, in Chapter 7 we shall discuss some cryptanalytic attacks on the short RSA public exponent e, whereas in Chapter 8 we shall discuss some attacks on the short RSA private exponent d. In Chapter 9, a completely different type of attacks, namely, the side-channel attacks on RSA, are discussed. Unlike the mathematical/algorithmic attacks in the previous chapters, side-channel attacks do not exploit the mathematical properties or weakness of the RSA algorithm/system itself, but exploit the hardware implementation issues of the system. In other words, these attacks are nothing to do with the RSA algorithm/system itself but have something to do with the hardware implementation of the RSA algorithm/system. Chapter 10, the final chapter, presents some quantum resistant, non-factoring based cryptographic systems as an alternative/replacement to RSA, such as lattice based and code-based cryptosystems, so that once RSA is proved to be insecure, there is an immediate replacement to the insecure RSA.

The book is self-contained and the materials presented in the book have been extensively classroom tested for various courses in Cryptography and Cryptanalysis at Aston and Coventry Universities in England, and the South China University of Technology and Nankai University in China. Many parts of the materials in the book have also been presented in seminars in various universities around the world. Hence, the book is suitable *either* as a research reference for public-key cryptology in general and for RSA cryptology in particular, *or* as a graduate text in the field.

Acknowledgments

The author would like to thank Prof Sushil Jajodia of George Mason University, USA, Prof Glyn James and Dr Anne James of Coventry University, UK, Prof Stephen Cook of the University of Toronto, Canada, and Prof Richard Brent of Oxford University and Australian National University for their encouragement, support and help during the writing of the book. Special thanks must also be given to Susan Lagerstrom-Fife and Sharon Palleschi, the editors at Springer in Boston, USA, for their encouragement, support and help.

Parts of book were written while the author visited the following three places: the Department of Computer Science at the University of Toronto (UT) in March-April 2005, hosted by Prof Stephen Cook and supported by UT and the Royal Society London, the Mathematical Science Institute at the Australian National University (ANU) in October-November 2006, hosted by Prof Richard Brent and supported by ANU and the Royal Society London, and the Department of Mathematics at the Massachusetts Institute of Technology (MIT) in July-Sept 2007, hosted by Prof Michael Sipser and supported by MIT.

Special thanks must also be given to Prof Glyn James at Coventry University and Prof Richard Brent at Oxford University and Australian National University for reading the whole manuscript of the book, to Prof Brain Scotney at Ulster University for his constant encouragement during the writing of the book, and to Prof Michael Sipser for inviting me to visit and work at MIT where the book was finally completed.

The struggle between code-makers and code-breakers is endless. The struggle between attacks and anti-attacks on RSA is also endless as soon as RSA is till in use. New ideas and new attacks on RSA may be conceived and invented anytime. So comments, corrections and suggestions on the book, and new ideas and news attacks on RSA are particularly very welcome from the readers, and can be sent to any one of my following three email addresses: song.yan@beds.ac.uk, syan@math.mit.edu, or syan@cs.toronto.edu, so that I can incorporate them into a future edition of the book. Thank you for your help in advance.

CAMBRIDGE, MASSACHUSETTS, AUGUST 2007 S. Y. Y.

Notation

All notation should be as simple as the nature of the operations to which it is applied.

CHARLES BABBAGE (1791–1871)
English Mathematician, Philosopher, Mechanical Engineer and Proto-Computer Scientist

Notation	**Explanation**
$\mathbb{N}$ or $\mathbb{Z}^+$	Set of natural numbers or positive integers: $\mathbb{N} = \mathbb{Z}^+ = \{1, 2, 3, \cdots\}$
$\mathbb{Z}$	Set of integers: $\mathbb{Z} = \{0, \pm n : n \in \mathbb{N}\}$
$\mathbb{Z}_{>1}$	Set of positive integers greater than 1: $\mathbb{Z}_{>1} = \{n : n \in \mathbb{Z} \text{ and } n > 1\}$
$\mathbb{Q}$	Set of rational numbers: $\mathbb{Q} = \left\{\frac{a}{b} : a, b \in \mathbb{Z} \text{ and } b \neq 0\right\}$
$\mathbb{R}$	Set of real numbers: $\mathbb{R} = \{n + 0.d_1 d_2 d_3 \cdots : n \in \mathbb{Z},\ d_i \in \{0, 1, 2, \cdots, 9\}$ and no infinite sequence of 9's appears$\}$
$\mathbb{C}$	Set of complex numbers: $\mathbb{C} = \{a + bi : a, b \in \mathbb{R} \text{ and } i = \sqrt{-1}\}$
$\mathbb{Z}_N$ or $\mathbb{Z}/N\mathbb{Z}$	Residue classes modulo N: $\mathbb{Z}_N = \mathbb{Z}/N\mathbb{Z} = \{0, 1, 2, \cdots, N-1\}$. Ring of integers. Field if N is prime
$\mathbb{Z}_N[x]$	Set (ring) of polynomials with integer coefficients, modulo N
$\mathbb{Z}[x]$	Set (ring) of polynomials with integer coefficients

$\mathbb{Z}_N^*$	Multiplicative group: $\mathbb{Z}_N^* = \{a \in \mathbb{Z}_N : \gcd(a, N) = 1\}$.
$\#(\mathbb{Z}_N^*)$ or $\|\mathbb{Z}_N^*\|$	Order of the multiplicative group
$\mathbb{F}_p$ or $\mathbb{Z}_p$	Finite field with p elements, where p is a prime
$\mathbb{F}_q$	Finite field with $q = p^k$ a prime power
$f(x)$	Function of x
f^{-1}	Inverse of f
$f(x) \sim g(x)$	$f(x)$ and $g(x)$ are asymptotically equal
$\binom{n}{i}$	Binomial coefficient: $\binom{n}{i} = \frac{n(n-1)(n-2)\cdots(n-i+1)}{i!}$
$\int$	Integration
$\mathrm{Li}(x)$	Logarithmic integral: $\mathrm{Li}(x) = \int_2^x \frac{\mathrm{d}t}{\ln t}$
$\sum_{i=1}^{n} x_i$	Sum: $x_1 + x_2 + \cdots + x_n$
$\prod_{i=1}^{n} x_i$	Product: $x_1 x_2 \cdots x_n$
x^k	x to the power k
kP	$kP = \underbrace{P \oplus P \oplus \cdots \oplus P}_{k \text{ summands}}$, where P is a point (x, y) on an elliptic curve $E: y^2 = x^3 + ax + b$
$\mathcal{O}_E$	Point at infinity on an elliptic curve E
$\log_b x$	Logarithm of x to the base b $(b \neq 1)$: $x = b^{\log_b x}$
$\log x$	Binary logarithm: $\log_2 x$
$\ln x$	Natural logarithm: $\log_e x$, $e = \sum_{n \geq 0} \frac{1}{n!} \approx 2.7182818$
$\exp(x)$	Exponential of x: $e^x = \sum_{n \geq 0} \frac{x^n}{n!}$
$a \mid b$	a divides b
$a \nmid b$	a does not divide b
$\gcd(a, b)$	Greatest common divisor of (a, b)
$\mathrm{lcm}(a, b)$	Least common multiple of (a, b)
$\lfloor x \rfloor$ or $[x]$	Greatest integer less than or equal to x
$\lceil x \rceil$	Least integer greater than or equal to x

$x \bmod \mathrm{N}$	Remainder: $x - N\left\lfloor \dfrac{x}{N} \right\rfloor$
$x = y \bmod N$	x is equal to y reduced to modulo N
$x \equiv y \pmod{\mathrm{N}}$	x is congruent to y modulo N
$x \not\equiv y \pmod{\mathrm{N}}$	x is not congruent to y modulo N
$x^k \bmod N$	x to the power k modulo N
$kP \bmod N$	kP modulo N, with P a point on elliptic curve E
$\mathrm{ord}_N(a)$	Order of an integer a modulo N; also denoted by $\mathrm{order}(a, N)$
$\mathrm{ind}_{g,N}(a)$	Index of a to the base g modulo N
$\log_g a \bmod N$	Discrete logarithm of a to the base g modulo N: $\log_g a \bmod N = \mathrm{ind}_{g,N}(a)$
$\pi(x)$	Prime counting function: $\pi(x) = \sum_{\substack{p \le x \\ p \text{ prime}}} 1$
$\tau(N)$	Number of (positive) divisors of N: $\tau(N) = \sum_{d \mid N} 1$
$\sigma(N)$	Sum of (positive) divisors of N: $\sigma(N) = \sum_{d \mid N} d$
$\phi(N)$	Euler's totient function: $\phi(N) = \sum_{\substack{0 \le k < N \\ \gcd(k,N)=1}} 1$
$\lambda(N)$	Carmichael's function: $\lambda(N) = \mathrm{lcm}\,(\lambda(p_1^{\alpha_1}), \lambda(p_2^{\alpha_2}), \cdots, \lambda(p_k^{\alpha_k}))$ if $N = \prod_{i=1}^{k} p_i^{\alpha_i}$
$\zeta(s)$	Riemann zeta-function: $\zeta(s) = \prod_{n=1}^{\infty} n^{-s}$, where $s = \sigma + it$, with $\sigma, t \in \mathbb{R}$ and $i = \sqrt{-1}$
$\left(\dfrac{a}{p}\right)$	Legendre symbol, where p is prime
$\left(\dfrac{a}{N}\right)$	Jacobi symbol, where n is composite
Q_N	Set of all quadratic residues of N
$\overline{Q}_N$	Set of all quadratic nonresidues of N
J_N	$J_N = \left\{ a \in \mathbb{Z}_N^* : \left(\dfrac{a}{N}\right) = 1 \right\}$
$\tilde{Q}_N$	Set of all pseudosquares of N: $\tilde{Q}_N = J_N - Q_N$
$\sim$	Asymptotic equality
$\approx$	Approximate equality
∞	Infinity

$\Longrightarrow$	Implication
$\Longleftrightarrow$	Equivalence
$\Box$	Blank symbol; end of proof
$\sqcup$	Space
Prob	Probability measure
$\in$	Member of
$\subset$	Proper subset
$\subseteq$	Subset
$[q_0, q_1, q_2, \cdots, q_n]$	Finite simple continued fraction: $q_0 + \cfrac{1}{q_1 + \cfrac{1}{q_2 + \cfrac{1}{\ddots\, q_{n-1} + \cfrac{1}{q_n}}}}$
$[q_0, q_1, q_2, q_3, \cdots]$	Infinite simple continued fraction: $q_0 + \cfrac{1}{q_1 + \cfrac{1}{q_2 + \cfrac{1}{q_3 + \cfrac{1}{\ddots}}}}$
$C_k = \dfrac{P_k}{Q_k}$	k-th convergent of a continued fraction
$\mathcal{P}$	Class of problems solvable in polynomial-time by a deterministic Turing machine
$\mathcal{NP}$	Class of problems solvable in polynomial-time by a nondeterministic Turing machine
$\mathcal{ZPP}$	Class of problems solvable in polynomial-time by a random Turing machine with zero errors
$\mathcal{RP}$	Class of problems solvable in polynomial-time by a random Turing machine with one-sided errors
$\mathcal{BPP}$	Class of problems solvable in in polynomial-time by a random Turing machine with two-sided errors
co-$\mathcal{RP}$	$\mathcal{RP}$-complete problems
co-$\mathcal{NP}$	$\mathcal{NP}$-complete problems
$\mathcal{PS}$	$\mathcal{P}$ $\mathcal{S}$pace problems
$\mathcal{NPS}$	$\mathcal{NP}$ $\mathcal{S}$pace problems
CFRAC	Continued FRACtion method (for factoring)
ECM	Elliptic Curve Method (for factoring)

NFS	Number Field Sieve (for factoring)
QS/MPQS	Quadratic Sieve/Multiple Polynomial Quadratic Sieve (for factoring)
ECPP	Elliptic Curve Primality Proving
DES	Data Encryption Standard
AES	Advanced Encryption Standard
DSA	Digital Signature Algorithm
DSS	Digital Signature Standard
DHM	Diffie-Hellman-Merkle key-exchange
RSA	Rivest-Shamir-Adleman Encryption
RSAP	RSA Problem
IFP	Integer Factorization Problem
DLP	Discrete Logarithm Problem
ECDLP	Elliptic Curve Discrete Logarithm Problem
QRP	Quadratic Residuosity Problem
SQRTP	Modular Square Root Problem
RFP	k-th Root Finding Problem
LLL	Lenstra-Lenstra-Lovasz lattice reduction algorithm
SVP	Shortest Vector Problem
PNT	Prime Number Theory: $\pi(x) \sim x/\ln x$
WWW	World Wide Web
$\mathcal{M}$	Plaintext space
M	$M \in \mathcal{M}$ Plaintext
$\mathcal{C}$	Ciphertext space
C	$C \in \mathcal{C}$ Ciphertext
e_k	Encryption key
d_k	Decryption key
$E_{e_k}(M)$	Encryption $C = E_{e_k}(M)$
$D_{d_k}(C)$	Decryption $M = D_{d_k}(C)$
e	RSA encryption exponent
d	RSA decryption exponent
$E_e(M)$	RSA encryption $C = E_e(M) \equiv M^e \pmod N$
$D_d(C)$	RSA decryption $M = D_d(C) \equiv C^d \pmod N$
$M \mapsto M^e \bmod N$	RSA function

$M^e \mapsto M \bmod N$	Inverse of the RSA function
$\{e, N, C\} \to \{M\}$	Given $e \equiv 1/e \pmod{\phi(N)}$, $N = pq$, and $C \equiv M^e \pmod{N}$, find $M \equiv C^d \pmod{N}\}$
$A \xrightarrow[\text{conditions/explainations}]{\text{conditions/explainations}} B$	B can be found from A under certain conditions
$A \xrightarrow[\text{hard}]{\text{find}} B$	B is hard to find from A
$A \xrightarrow[\text{easy}]{\text{find}} B$	B is easy to find from A
$\text{A} \overset{\mathcal{P}}{\Longrightarrow} \text{B}$	Given A, B can be solved in polynomial-time
$\text{A} \overset{\mathcal{P}}{\Longleftrightarrow} \text{B}$	A and B are deterministic polynomial-time equivalent
$\mathcal{O}(N)$	Upper bound: $f(N) = \mathcal{O}(g(N))$ if there exists *some* constant $c > 0$ such that $f(N) \le c \cdot g(N)$
$\mathcal{O}(N^k)$	Polynomial-time complexity measured in terms of arithmetic operations, where $k > 0$ is a constant
$\mathcal{O}\left((\log N)^k\right)$	Polynomial-time complexity measured in terms of bit operations, where $k > 0$ is a constant
$\mathcal{O}\left((\log N)^{c \log N}\right)$	Superpolynomial complexity, where $c > 0$ is a constant
$\mathcal{O}\left(\exp\left(c\sqrt{\log N \log\log N}\ \right)\right)$	Subexponential complexity: $\mathcal{O}\left(\exp\left(c\sqrt{\log N \log\log N}\ \right)\right) = \mathcal{O}\left(N^{c\sqrt{\log\log N/\log N}}\right)$, where c is a constant;
$\mathcal{O}(\exp(x))$	Exponential complexity, sometimes denoted by $\mathcal{O}(e^x)$
$\mathcal{O}(N^\epsilon)$	Exponential complexity measured in terms of bit operations: $\mathcal{O}(N^\epsilon) = \mathcal{O}\left(2^{\epsilon \log N}\right)$, $\epsilon > 0$ is constant

1. Computational/Mathematical Preliminaries

The problem of good cipher design is essentially one of finding difficult problems, subject to certain other conditions.

CLAUDE E. SHANNON (1916–2001)
Father of Information Theory

1.1 Introduction

Problems, particularly *intractable number-theoretic problems*, are the main concern of this book, as these problems are intimately connected with *public-key cryptography*, particularly *RSA public-key cryptography.* In fact, the security of any public-key cryptosystems relies, in one way or another, on the intractability of some sort of number-theoretic problems, or more generally, mathematical problems such as problems in arithmetic algebraic geometry.

From a modern computer science point of view, problems may be classified into two categories: *solvable or computable problems*, and *unsolvable or uncomputable problems.* The solvable problems can further be divided into two subcategories (see Figure 1.1): tractable or feasible problems, and intractable or infeasible problems. Informally, a problem is solvable if it can be solved by a Turing machine, or otherwise, it is unsolvable. A problem is tractable (easy) if it can be solved in polynomial-time by a deterministic Turing machine, or otherwise, it is intractable (hard).

It should be noted that problems solvable in polynomial-time on a Turing machine are exactly the same as the problems solvable in polynomial-time on a typical computer. This implies that Turing machines (or algorithms, effective procedures) and computers are equivalent concepts, and we shall use them interchangeably. It should be also noted that the classification of problems solvable in polynomial-time (i.e., tractable or feasible) and unsolvable in polynomial-time e.g., solvable in exponential-time (i.e., intractable or feasible) is fundamental in computer science. However, there is no clear

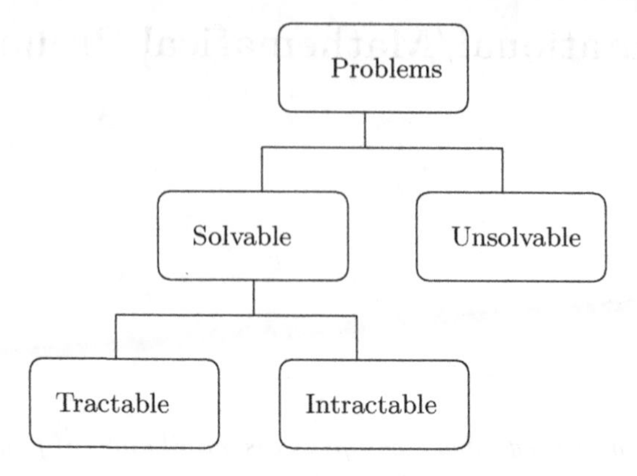

Figure 1.1. Taxonomy of Problems

cut between tractable and intractable problems. To delineate the boundary between the two types of problems is, in fact, the famous $\mathcal{P}$ versus $\mathcal{NP}$ Problem, a one-million US dollar millennium prize problem, offered by the Clay Mathematical Institute in Boston [73].

In public-key cryptography, particularly RSA public-key cryptography, we are interested in both tractable and intractable number-theoretic problems and their computations. For example, for an authorized user, the encryption and decryption should be all intractable for him, however, for an unauthorized user, the decryption must be intractable although encryption is easy for everyone. Figure 1.2 shows some tractable and intractable problems in number theory. So, in this chapter, we shall study some efficient algorithms for tractable number-theoretic problems, such as

(1) Euclid's algorithm for
- computing the greatest common divisor of two integers a and b, $\gcd(a, b)$,
- solving the linear congruence

$$ax \equiv 1 \pmod{N}, \quad \text{or} \quad ax \equiv b \pmod{N},$$

- solving the linear Diophantine equation

$$ax + by = c.$$

(2) Continued fraction algorithm for representing a rational number a/b into its continued fraction expansion,

$$a/b = q_0 + \cfrac{1}{q_1 + \cfrac{1}{q_2 + \cfrac{1}{\ddots\, q_{n-1} + \cfrac{1}{q_n}}}}.$$

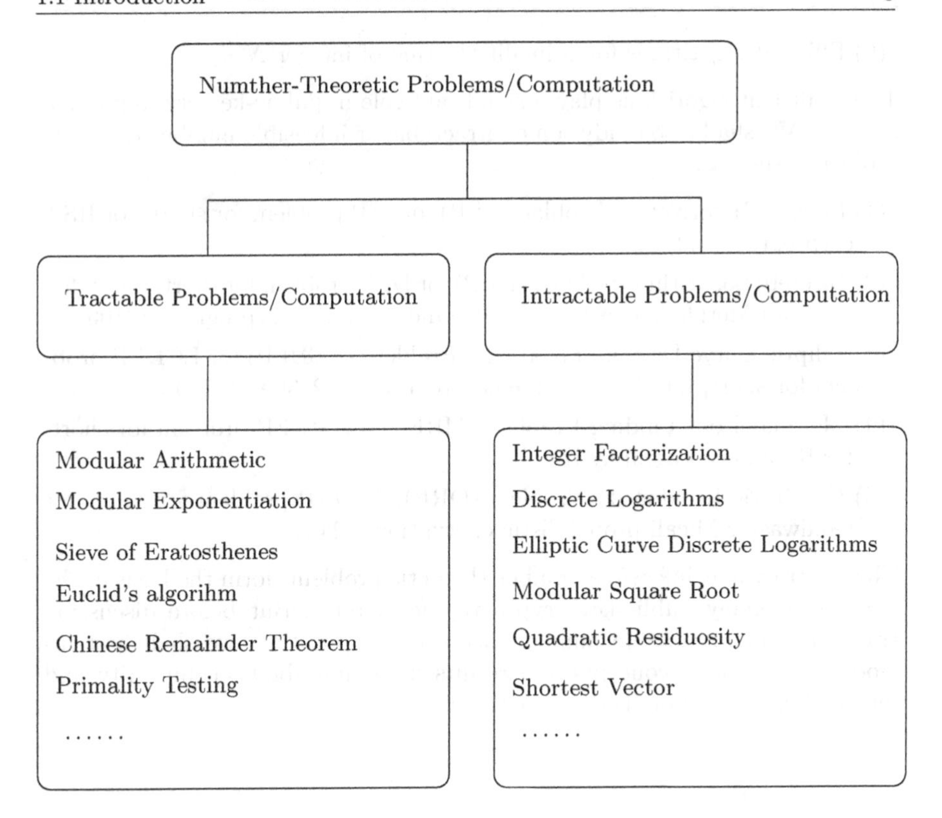

Figure 1.2. Taxonomy of Problems in Number Theory

(3) Chinese Remainder Theorem for solving the system of congruences,

$$\begin{cases} x \equiv a_1 \pmod{m_1}, \\ x \equiv a_2 \pmod{m_2}, \\ \quad \cdots\cdots \\ \quad \cdots\cdots \\ x \equiv a_n \pmod{m_n}. \end{cases}$$

(4) Repeated squaring algorithm for fast modular exponentiation,

$$y \equiv y^k \pmod{N}.$$

(5) Repeated doubling algorithm for fast group operations on elliptic curves,

$$Q \equiv kP \pmod{N},$$

where P and Q are points on the elliptic curve $E : y^2 = x^3 + ax + b$ over $\mathbb{Z}_N$.

(6) Efficient algorithms for primality testing of integer N.

These efficient algorithms play a significant role in public-key cryptographic systems. We shall also study some intractable or infeasible number-theoretic problems, such as:

(1) Integer Factorization Problem (IFP), or IFP problem for short, for RSA cryptography [262].
(2) Discrete Logarithm Problem (DLP), or DLP problem for short, for Diffie-Hellman-Merkle key-exchange [101] and ElGamal cryptography [106].
(3) Elliptic Curve Discrete Logarithm Problem (ECDLP), or ECDLP problem for short, for elliptic curve cryptography ([213] and [170]).
(4) Modular Root Finding Problem (MRFP), or MRFP problem for short, for Rabin cryptography [252].
(5) Quadratic Residuosity Problem (QRP), or QRP problem for short, for Goldwasser-Micali probabilistic encryption [124].

These intractable/infeasible number-theoretic problems form the basis of the security of many public-key cryptographic systems. But before discussing tractable and intractable number-theoretic computation, we shall first introduce some basic concepts and results in computability, complexity and intractability based on Turing machines.

1.2 Computability, Complexity and Intractability

Turing Machines

The theory of Turing machines was first proposed and studied by the great English logician and mathematician Alan Turing in his seminal paper [316] published in 1936 (see Figure 1.3 for the first page the paper):

230 A. M. TURING [Nov. 12,

ON COMPUTABLE NUMBERS, WITH AN APPLICATION TO THE ENTSCHEIDUNGSPROBLEM

By A. M. TURING.

[Received 28 May, 1936.—Read 12 November, 1936.]

The "computable" numbers may be described briefly as the real numbers whose expressions as a decimal are calculable by finite means. Although the subject of this paper is ostensibly the computable *numbers*, it is almost equally easy to define and investigate computable functions of an integral variable or a real or computable variable, computable predicates, and so forth. The fundamental problems involved are, however, the same in each case, and I have chosen the computable numbers for explicit treatment as involving the least cumbrous technique. I hope shortly to give an account of the relations of the computable numbers, functions, and so forth to one another. This will include a development of the theory of functions of a real variable expressed in terms of computable numbers. According to my definition, a number is computable if its decimal can be written down by a machine.

In §§ 9, 10 I give some arguments with the intention of showing that the computable numbers include all numbers which could naturally be regarded as computable. In particular, I show that certain large classes of numbers are computable. They include, for instance, the real parts of all algebraic numbers, the real parts of the zeros of the Bessel functions, the numbers π, e, etc. The computable numbers do not, however, include all definable numbers, and an example is given of a definable number which is not computable.

Although the class of computable numbers is so great, and in many ways similar to the class of real numbers, it is nevertheless enumerable. In § 8 I examine certain arguments which would seem to prove the contrary. By the correct application of one of these arguments, conclusions are reached which are superficially similar to those of Gödel†. These results

† Gödel, "Über formal unentscheidbare Sätze der Principia Mathematica und verwandter Systeme, I", *Monatshefte Math. Phys.*, 38 (1931), 173–198.

Figure 1.3. The First Page of Turing's 1936 Paper

First of all we present a formal definition of Turing machines.

Definition 1.2.1. A standard multitape *Turing machine*, M (see Figure 1.4), is an algebraic system defined by

$$M = (Q, \Sigma, \Gamma, \delta, q_0, \Box, F) \tag{1.1}$$

where

(1) Q is a finite set of *internal states*;

(2) Σ is a finite set of symbols called the *input alphabet*. We assume that $\Sigma \subseteq \Gamma - \{\Box\}$;

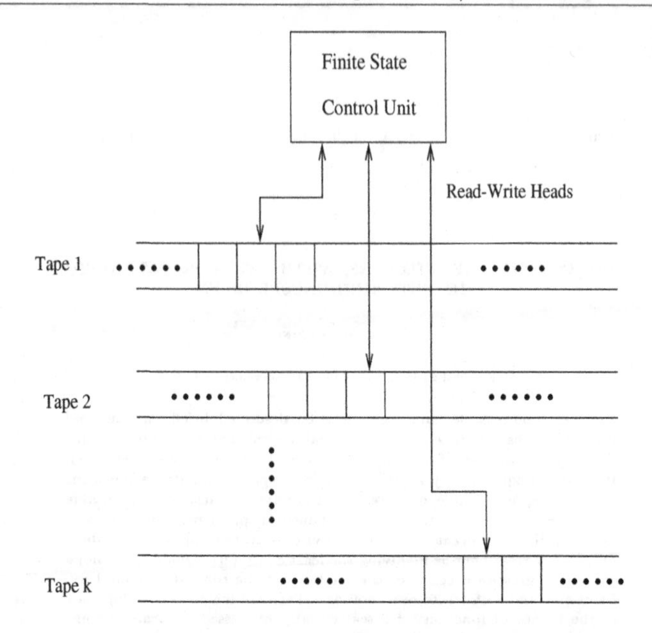

Figure 1.4. A k-tape ($k \geq 1$) Turing Machine

(3) Γ is a finite set of symbols called the *tape alphabet*;

(4) δ is the transition function, which is defined by

(i) if M is a deterministic Turing machine (DTM), then

$$\delta: \; Q \times \Gamma^k \to Q \times \Gamma^k \times \{L, R\}^k \tag{1.2}$$

(ii) if M is a nondeterministic Turing machine (NDTM), then

$$\delta : Q \times \Gamma^k \to 2^{Q \times \Gamma^k \times \{L,R\}^k} \tag{1.3}$$

where L and R specify the movement of the read-write head *left* or *right*. When $k = 1$, it is just a standard one-tape Turing machine;

(5) $\square \in \Gamma$ is a special symbol called the *blank*;

(6) $q_0 \in Q$ is the *initial state*;

(7) $F \subseteq Q$ is the set of *final states*.

Thus, Turing machines provide us with the simplest possible abstract model of computation for modern digital (even quantum) computers.

Example 1.2.1. Given two positive integers x and y design a Turing machine that computes $x + y$.

First, we have to choose some convention for representing positive integers. For simplicity, we will use unary notation in which any positive integer

x is represented by $w(x) \in \{1\}^+$, such that $|w(x)| = x$. Thus in this notation, 4 will be represented by 1111. We must also decide how x and y are placed on the tape initially and how their sum is to appear at the end of the computation. It is assumed that $w(x)$ and $w(y)$ are on the tape in unary notation, separated by a single 0, with the read-write head on the leftmost symbol of $w(x)$. After the computation, $w(x+y)$ will be on the tape followed by a single 0, and the read-write head will be positioned at the left end of the result. We therefore want to design a Turing machine for performing the computation

$$q_0 w(x)0w(y) \overset{*}{\vdash} q_f w(x+y)0,$$

where $q_f \in F$ is a final state. Constructing a program for this is relatively simple. All we need to do is to move the separating 0 to the right end of $w(y)$, so that the addition amounts to nothing more than the coalition of the two strings. To achieve this, we construct

$$M = (Q, \Sigma, \Gamma, \delta, q_0, \Box, F),$$

with

$$\begin{aligned}
Q &= \{q_0, q_1, q_2, q_3, q_4\}, \\
F &= \{q_4\}, \\
\delta(q_0, 1) &= (q_0, 1, R), \\
\delta(q_0, 0) &= (q_1, 1, R), \\
\delta(q_1, 1) &= (q_1, 1, R), \\
\delta(q_1, \Box) &= (q_2, \Box, L), \\
\delta(q_2, 1) &= (q_3, 0, L), \\
\delta(q_3, 1) &= (q_3, 1, L),
\end{aligned}$$

Note that in moving the 0 right we temporarily create an extra 1, a fact that is remembered by putting the machine into state q_1. The transition $\delta(q_2, 1) = (q_3, 0, R)$ is needed to remove this at the end of the computation. This can be seen from the sequence of instantaneous descriptions for adding 111 to 11:

$$
\begin{aligned}
q_0 1110011 &\vdash 1 q_0 110011 \\
&\vdash 11 q_0 1011 \\
&\vdash 111 q_0 011 \\
&\vdash 1111 q_1 11 \\
&\vdash 11111 q_1 1 \\
&\vdash 111111 q_1 \\
&\vdash 11111 q_2 1 \\
&\vdash 1111 q_3 10 \\
&\overset{*}{\vdash} q_3 \square 111110 \\
&\vdash q_4 111110.
\end{aligned}
$$

Definition 1.2.2. A function is effectively *computable or uncomputable* if it can or cannot be computed by a Turing machine. A problem is solvable or unsolvable if it can or cannot be solved by a Turing machine. A langauge is acceptable or unacceptable if it can or cannot be accepted by a Turing machine.

The Church-Turing Thesis

Any *effectively* computable function can be computed by a Turing machine, and there is no effective procedure that a Turing machine cannot perform. This leads naturally to the following famous Church-Turing thesis, named after Alonzo Church (1903–1995) and Alan Turing (1912–1954):

> **The Church-Turing thesis**: Any effectively computable function can be computed by a Turing machine.

The Church-Turing thesis thus provides us with a powerful tool to distinguish what is computation and what is not computation, what function is computable and what function is not computable, and more generally, what computers can do and what computers cannot do.

It must be noted that the Church-Turing thesis is not a mathematical theorem, and hence it cannot be proved formally, since, to prove the Church-Turing thesis, we need to formalize what is effectively computable, which is impossible. However, many computational evidences support the thesis and in fact no counterexample has been found yet.

From a computer science and particularly a cryptographic point of view, we are not just interested in what computers can do, but in what computers can do efficiently. That is, in cryptography we are more interested in practical computable rather than just theoretical computable; this leads to the Cook-Karp Thesis.

Remark 1.2.1. Church in his famous 1936 paper [69] (see the first page of his paper in Figure 1.5) proposed the important concept of λ-definable and later in his book review [70] of Turing's 1936 paper, he said that all effective procedures are in fact Turing equivalent. This is what we call now the Church-Turing Thesis.

AN UNSOLVABLE PROBLEM OF ELEMENTARY NUMBER THEORY.[1]

By ALONZO CHURCH.

1. **Introduction.** There is a class of problems of elementary number theory which can be stated in the form that it is required to find an effectively calculable function f of n positive integers, such that $f(x_1, x_2, \cdots, x_n) = 2$ [2] is a necessary and sufficient condition for the truth of a certain proposition of elementary number theory involving $x_1, x_2, \cdots, x_n$ as free variables.

An example of such a problem is the problem to find a means of determining of any given positive integer n whether or not there exist positive integers x, y, z, such that $x^n + y^n = z^n$. For this may be interpreted, required to find an effectively calculable function f, such that $f(n)$ is equal to 2 if and only if there exist positive integers x, y, z, such that $x^n + y^n = z^n$. Clearly the condition that the function f be effectively calculable is an essential part of the problem, since without it the problem becomes trivial.

Another example of a problem of this class is, for instance, the problem of topology, to find a complete set of effectively calculable invariants of closed three-dimensional simplicial manifolds under homeomorphisms. This problem can be interpreted as a problem of elementary number theory in view of the fact that topological complexes are representable by matrices of incidence. In fact, as is well known, the property of a set of incidence matrices that it represent a closed three-dimensional manifold, and the property of two sets of incidence matrices that they represent homeomorphic complexes, can both be described in purely number-theoretic terms. If we enumerate, in a straightforward way, the sets of incidence matrices which represent closed three-dimensional manifolds, it will then be immediately provable that the problem under consideration (to find a complete set of effectively calculable invariants of closed three-dimensional manifolds) is equivalent to the problem, to find an effectively calculable function f of positive integers, such that $f(m, n)$ is equal to 2 if and only if the m-th set of incidence matrices and the n-th set of incidence matrices in the enumeration represent homeomorphic complexes.

Other examples will readily occur to the reader.

[1] Presented to the American Mathematical Society, April 19, 1935.

[2] The selection of the particular positive integer 2 instead of some other is, of course, accidental and non-essential.

345

Figure 1.5. The First Page of Church's 1935 Paper

Complexity Classes

We are now in a position to present a formal definition of some common complexity classes based computations on Turing machines. First, we need a definition for probabilistic or randomized Turing machines.

Definition 1.2.3. A *probabilistic Turing machine* is a type of nondeterministic Turing machine with distinct states called *coin-tossing states*. For each coin-tossing state, the finite control unit specifies two possible legal next states. The computation of a probabilistic Turing machine is deterministic except that in coin-tossing states the machine tosses an unbiased coin to decide between the two *possible legal* next states.

A probabilistic Turing machine can be viewed as a *randomized Turing machine* [145], as described in Figure 1.6. The first tape, holding input, is

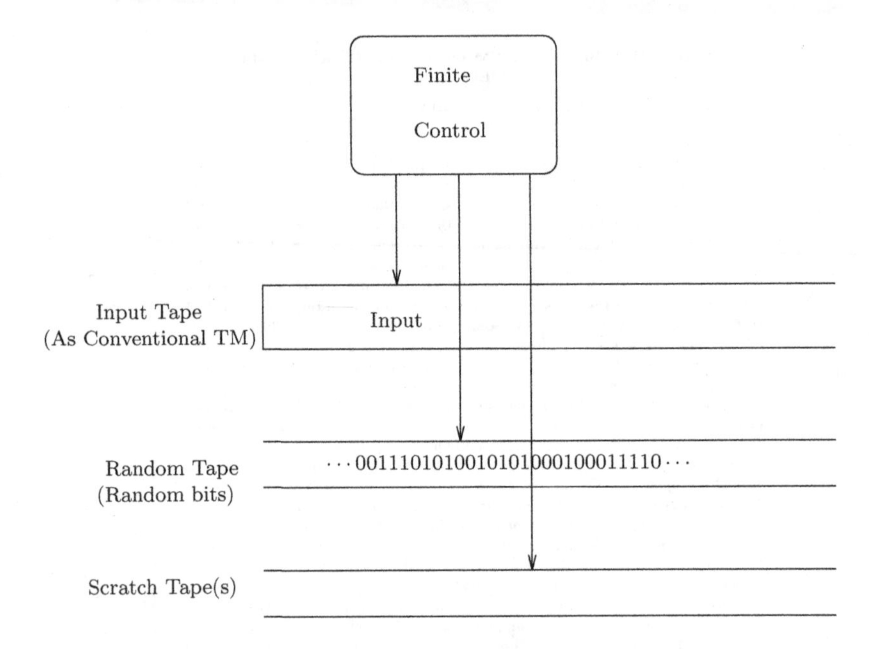

Figure 1.6. A k-tape ($k \geq 1$) Turing Machine

just the same as conventional multitape Turing machine. The second tape is referred to as random tape, containing randomly and independently chosen bits, with probability 1/2 of a 0 and the same probability 1/2 of a 1. The third and subsequent tapes are used, if needed, as scratch tapes by the Turing machine.

Definition 1.2.4. $\mathcal{P}$ is the class of problems solvable in polynomial-time by a deterministic Turing machine (DTM). Problems in this class are classified to be tractable (feasible) and easy to solve on a computer. For example, additions of any two integers, no matter how big they are, can be performed in polynomial-time, and hence it is in $\mathcal{P}$.

Definition 1.2.5. $\mathcal{NP}$ is the class of problems solvable in polynomial-time on a nondeterministic Turing machine (NDTM). Problems in this class are classified to be intractable (infeasible) and hard to solve on a computer. For example, the Traveling Salesman Problem (TSP) is in $\mathcal{NP}$, and hence it is hard to solve.

In terms of formal languages, we may also say that $\mathcal{P}$ is the class of languages where the membership in the class can be decided in polynomial-time, whereas $\mathcal{NP}$ is the class of languages where the membership in the class can be verified in polynomial-time [304]. It seems that the power of polynomial-time verifiable is greater than that of polynomial-time decidable, but no proof has been given to support this statement (see Figure 1.7). The question of whether or not $\mathcal{P} = \mathcal{NP}$ is one of the greatest unsolved problems in computer science and mathematics, and in fact it is one of the seven Millennium Prize Problems proposed by the Clay Mathematics Institute in Boston in 2000, each with one-million US dollars [73].

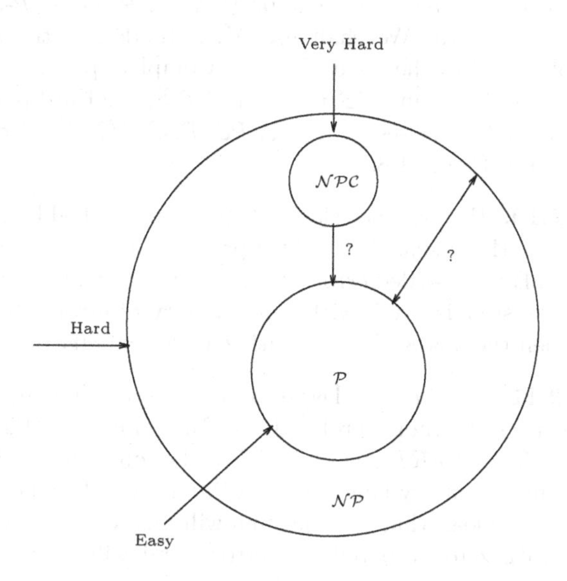

Figure 1.7. The $\mathcal{P}$ Versus $\mathcal{NP}$ Problem

Definition 1.2.6. $\mathcal{EXP}$ is the class of problems solvable by a deterministic Turing machine (DTM) in time bounded by 2^{n^i}.

Definition 1.2.7. A function f is polynomial-time computable if for any input w, $f(w)$ will halt on a Turing machine in polynomial-time. A language A is polynomial-time reducible to a langauge B, denoted by $A \leq_{\mathcal{P}} B$, if there exists a polynomial-time computable function such that for every input w,

$$w \in A \Longleftrightarrow f(w) \in B.$$

The function f is called the polynomial-time reduction of A to B.

Definition 1.2.8. A language/problem L is $\mathcal{NP}$-Completeness if it satisfies the following two conditions:

(1) $L \in \mathcal{NP}$,

(2) $\forall A \in \mathcal{NP}, \;\; A \leq_{\mathcal{P}} L$.

Definition 1.2.9. A problem D is $\mathcal{NP}$-hard if it satisfies the following condition:

$$\forall A \in \mathcal{NP}, \;\; A \leq_{\mathcal{P}} D$$

where D may be in $\mathcal{NP}$, or may not be in $\mathcal{NP}$. Thus, $\mathcal{NP}$-hard means *at least as hard as any $\mathcal{NP}$-problem*, although it might, in fact, be harder.

Similarly, one can define the class of problems of $\mathcal{P}$-Space, $\mathcal{P}$-Space Complete, and $\mathcal{P}$-Space Hard. We shall use $\mathcal{NPC}$ to denote the set of $\mathcal{NP}$-Complete problems, $\mathcal{PSC}$ the set of $\mathcal{P}$-Space Complete problems, $\mathcal{NPH}$ the set of $\mathcal{NP}$-hard problems, and $\mathcal{PSH}$ the set of $\mathcal{P}$-Space Hard problems. The relationships among the classes $\mathcal{P}$, $\mathcal{NP}$, $\mathcal{NPC}$, $\mathcal{PSC}$, $\mathcal{NPH}$, $\mathcal{PSH}$, and $\mathcal{EXP}$ may be described in Figure 1.8.

Definition 1.2.10. $\mathcal{RP}$ is the class of problems solvable in expected polynomial-time with *one-sided error* by a probabilistic (randomized) Turing machine (PTM). By "one-sided error" we mean that the machine will answer "yes" when the answer is "yes" with a probability of error $< 1/2$, and will answer "no" when the answer is "no" with zero probability of error.

Definition 1.2.11. $\mathcal{ZPP}$ is the class of problems solvable in expected polynomial time with *zero error* on a probabilistic Turing machine (PTM). It is defined by $\mathcal{ZPP} = \mathcal{RP} \cap$ co-$\mathcal{RP}$, where co-$\mathcal{RP}$ is the complement of $\mathcal{RP}$. where co-$\mathcal{RP}$ is the complementary language of $\mathcal{RP}$. i.e., co-$\mathcal{RP} = \{L : \overline{L} \in \mathcal{RP}\}$). By "zero error" we mean that the machine will answer "yes" when the answer is "yes" (with zero probability of error), and will answer "no" when the answer is "no" (also with zero probability of error). But note that the machine may also answer "?", which means that the machine does not know the answer is "yes" or "no". However, it is guaranteed that at most half of simulation cases the machine will answer "?". $\mathcal{ZPP}$ is usually referred to an *elite class*, because it also equals to the class of problems that can be solved by randomized algorithms that always give the correct answer and run in expected polynomial time.

Definition 1.2.12. $\mathcal{BPP}$ is the class of problems solvable in expected polynomial-time with *two sided error* on a probabilistic Turing machine (PTM), in which the answer always has probability at least $\frac{1}{2} + \delta$, for some fixed $\delta > 0$ of being correct. The "$\mathcal{B}$" in $\mathcal{BPP}$ stands for "bounded away the error probability from $\frac{1}{2}$"; for example, the error probability could be $\frac{1}{3}$.

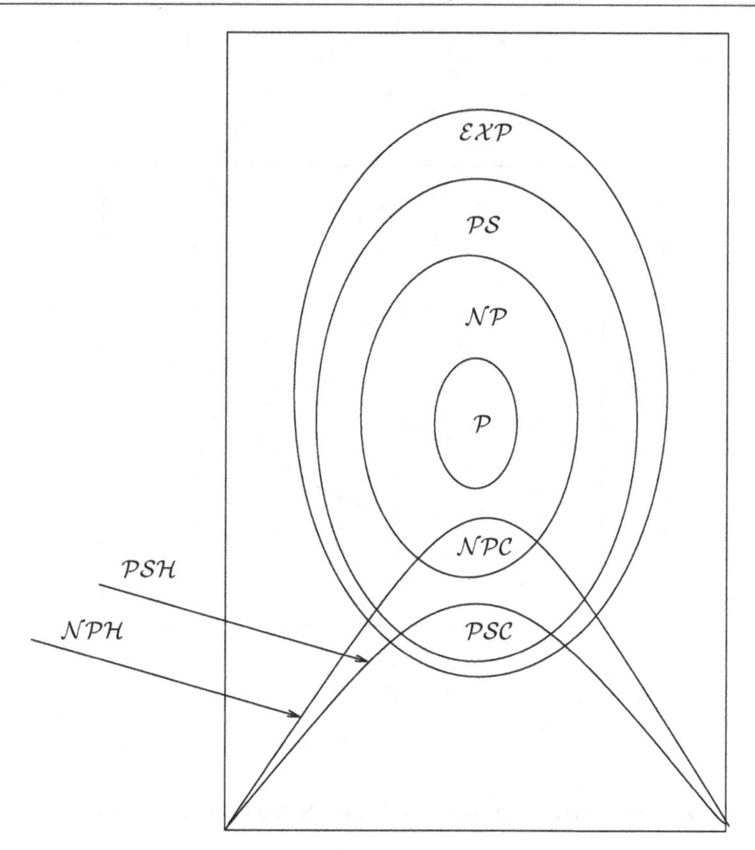

Figure 1.8. Conjectured Relationships Among Classes $\mathcal{P}$, $\mathcal{NP}$ and $\mathcal{NPC}$, etc.

The space complexity classes $\mathcal{P}$-SPACE and $\mathcal{NP}$-SPACE can be defined analogously as $\mathcal{P}$ and $\mathcal{NP}$. It is clear that a time class is included in the corresponding space class since one unit is needed to the space by one square. Although it is not known whether or not $\mathcal{P} = \mathcal{NP}$, it is known that $\mathcal{P}$-SPACE $= \mathcal{NP}$-SPACE. It is generally believed that

$$\mathcal{P} \subseteq \mathcal{ZPP} \subseteq \mathcal{RP} \subseteq \begin{pmatrix} \mathcal{BPP} \\ \mathcal{NP} \end{pmatrix} \subseteq \mathcal{P}\text{-SPACE} \subseteq \mathcal{EXP}.$$

Besides the proper inclusion $\mathcal{P} \subset \mathcal{EXP}$, it is not known whether any of the other inclusions in the above hierarchy is proper. Note that the relationship of $\mathcal{BPP}$ and $\mathcal{NP}$ is not known, although it is believed that $\mathcal{NP} \not\subseteq \mathcal{BPP}$. Figure 1.9 shows the relationships among the various common complexity classes.

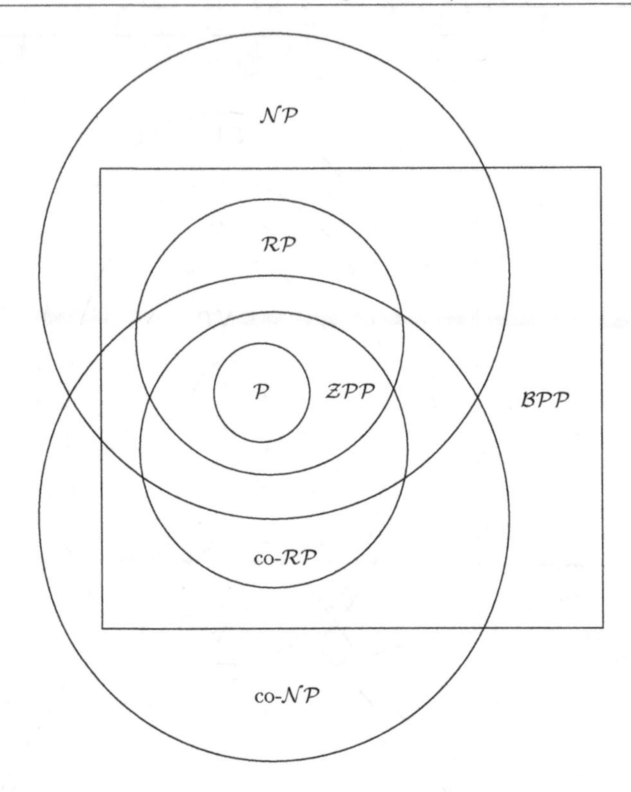

Figure 1.9. Conjectured Relationships Among Some Common Complexity Classes

The Cook-Karp Thesis

It is widely believed, although no proof has been given, that problems in $\mathcal{P}$ are computationally tractable, whereas problems not in (beyond) $\mathcal{P}$ are computationally intractable. (see Figure 1.15 for more information). This is the famous *Cook-Karp thesis*, named after Stephen Cook (Figure 1.10 shows the first page of Cook's paper) and Richard Karp (Figure 1.11 shows the first page of Karp's paper):

> **The Cook-Karp thesis**. Any computationally tractable problem can be computed by a Turing machine in deterministic polynomial-time.

Thus, problems in $\mathcal{P}$ are tractable whereas problems in $\mathcal{NP}$ are intractable. However, there is not a clear cut between the two types of problems. This is exactly the $\mathcal{P}$ Versus $\mathcal{NP}$, mentioned earlier.

Compared to the Church-Turing Thesis, the Cook-Karp Thesis provides a step closer to practical computability and complexity, and hence the life after Cook and Karp is much easier, since there is no need to go all the way back to

The Complexity of Theorem-Proving Procedures

Stephen A. Cook

University of Toronto

Summary

It is shown that any recognition problem solved by a polynomial time-bounded nondeterministic Turing machine can be "reduced" to the problem of determining whether a given propositional formula is a tautology. Here "reduced" means, roughly speaking, that the first problem can be solved deterministically in polynomial time provided an oracle is available for solving the second. From this notion of reducible, polynomial degrees of difficulty are defined, and it is shown that the problem of determining tautologyhood has the same polynomial degree as the problem of determining whether the first of two given graphs is isomorphic to a subgraph of the second. Other examples are discussed. A method of measuring the complexity of proof procedures for the predicate calculus is introduced and discussed.

Throughout this paper, a set of strings means a set of strings on some fixed, large, finite alphabet Σ. This alphabet is large enough to include symbols for all sets described here. All Turing machines are deterministic recognition devices, unless the contrary is explicitly stated.

1. Tautologies and Polynomial Re-Reducibility.

Let us fix a formalism for the propositional calculus in which formulas are written as strings on Σ. Since we will require infinitely many proposition symbols (atoms), each such symbol will consist of a member of Σ followed by a number in binary notation to distinguish that symbol. Thus a formula of length n can only have about $n/\log n$ distinct function and predicate symbols. The logical connectives are & (and), $\vee$ (or), and $\neg$(not).

The set of tautologies (denoted by {tautologies}) is a certain recursive set of strings on this alphabet, and we are interested in the problem of finding a good lower bound on its possible recognition times. We provide no such lower bound here, but theorem 1 will give evidence that {tautologies} is a difficult set to recognize, since many apparently difficult problems can be reduced to determining tautologyhood. By reduced we mean, roughly speaking, that if tautologyhood could be decided instantly (by an "oracle") then these problems could be decided in polynomial time. In order to make this notion precise, we introduce query machines, which are like Turing machines with oracles in [1].

A query machine is a multitape Turing machine with a distinguished tape called the query tape, and three distinguished states called the query state, yes state, and no state, respectively. If M is a query machine and T is a set of strings, then a T-computation of M is a computation of M in which initially M is in the initial state and has an input string w on its input tape, and each time M assumes the query state there is a string u on the query tape, and the next state M assumes is the yes state if $u \in T$ and the no state if $u \notin T$. We think of an "oracle", which knows T, placing M in the yes state or no state.

Definition

A set S of strings is P-reducible (P for polynomial) to a set T of strings iff there is some query machine M and a polynomial $Q(n)$ such that for each input string w, the T-computation of M with input w halts within $Q(|w|)$ steps ($|w|$ is the length of w), and ends in an accepting state iff $w \in S$.

It is not hard to see that P-reducibility is a transitive relation. Thus the relation E on

-151-

Figure 1.10. The First Page of Cook's Paper

Church and Turing. Again, Cook-Karp Thesis is not a mathematical theorem and hence cannot be proved mathematically, however evidences support the thesis.

1.3 Efficient Number-Theoretic Algorithms

In this section, we study some well-known and widely used efficient number-theoretic algorithms, including Euclid's algorithm, continued fraction algo-

REDUCIBILITY AMONG COMBINATORIAL PROBLEMS†

Richard M. Karp

University of California at Berkeley

Abstract: A large class of computational problems involve the determination of properties of graphs, digraphs, integers, arrays of integers, finite families of finite sets, boolean formulas and elements of other countable domains. Through simple encodings from such domains into the set of words over a finite alphabet these problems can be converted into language recognition problems, and we can inquire into their computational complexity. It is reasonable to consider such a problem satisfactorily solved when an algorithm for its solution is found which terminates within a number of steps bounded by a polynomial in the length of the input. We show that a large number of classic unsolved problems of covering, matching, packing, routing, assignment and sequencing are equivalent, in the sense that either each of them possesses a polynomial-bounded algorithm or none of them does.

1. INTRODUCTION

All the general methods presently known for computing the chromatic number of a graph, deciding whether a graph has a Hamilton circuit, or solving a system of linear inequalities in which the variables are constrained to be 0 or 1, require a combinatorial search for which the worst case time requirement grows exponentially with the length of the input. In this paper we give theorems which strongly suggest, but do not imply, that these problems, as well as many others, will remain intractable perpetually.

†This research was partially supported by National Science Foundation Grant GJ-474.

85

Figure 1.11. The First Page of Karp's Paper

rithm, modular exponentiations, the Chinese Remainder Theorem, point additions of elliptic curves, and primality testing algorithms.

Euclid's Algorithm

Our first and also the most important efficient algorithm in number theory and cryptography is Euclid's algorithm for computing $\gcd(a, b)$, and its extension for solving the linear congruence

$$ax \equiv 1 \pmod{n}$$

or

$$ax \equiv b \pmod{n}$$

and the linear Diophantine equation

$$ax - by = c.$$

First notice that $ax \equiv 1 \pmod{n}$ is a special case of $ax \equiv b \pmod{n}$. Notice again that $ax \equiv b \pmod{n}$ is equivalent to $ax - by = c$, since, e.g.,

$$ax \equiv b \pmod{n} \Longrightarrow ax - ny = b.$$

Thus, it suffices to only consider the solution to $ax + by = c$.

Theorem 1.3.1. Let a, b, c be integers with not both a and b equal to 0, and let $d = \gcd(a, b)$. If $d \nmid c$, then the linear Diophantine equation

$$ax + by = c \tag{1.4}$$

has no integer solution. The equation has an integer solution in x and y if and only if $d \mid c$. Moreover, if (x_0, y_0) is an integer solution of the equation, then the general solution of the equation is

$$(x, y) = \left(x_0 + \frac{b}{d} \cdot t,\ y_0 - \frac{a}{d} \cdot t\right) \tag{1.5}$$

where $t \in \mathbb{Z}$ is an integer parameter.

Proof. Assume that x and y are integers such that $ax + by = c$. Since $d \mid a$ and $d \mid b$, $d \mid c$. Hence, if $d \nmid c$, there is no integer solutions of the equation.

Now suppose $d \mid c$. There is an integer k such that $c = kd$. Since d is a sum of multiples of a and b, we may write

$$am + bn = d.$$

Multiplying this equation by k, we get

$$a(mk) + b(nk) = dk = c$$

so that $x = mk$ and $y = nk$ is a solution.

For the "only if" part, suppose x_0 and y_0 is a solution of the equation. Then

$$ax_0 + by_0 = c.$$

Since $d \mid a$ and $d \mid b$, then $d \mid c$. □

Theorem 1.3.2 (Division Theorem). For any integer a and any positive integer b, there exist unique integers q and r such that

$$a = bq + r, \quad 0 \le r < b, \tag{1.6}$$

where a is called the *dividend*, q the *quotient*, and r the *remainder*. If $b \nmid a$, then r satisfies the stronger inequalities $0 < r < a$.

Proof. Consider the arithmetic progression

$$\cdots, -3b, -2b, -b, 0, b, 2b, 3b, \cdots$$

then there must be an integer q such that

$$qb \leq a \leq (q+1)b.$$

Let $a - qb = r$, then $a = bq + r$ with $0 \leq r < b$. To prove the uniqueness of q and r, suppose there is another pair q_1 and r_1 satisfying the same condition in (1.6), then

$$a = bq_1 + r_1, \quad 0 \leq r_1 < b.$$

We first show that $r_1 = r$. For if not, we may presume that $r < r_1$, so that $0 < r_1 - r < b$, and then we see that $b(q - q_1) = r_1 - r$, and so $b \mid (r_1 - r)$, which is impossible. Hence, $r = r_1$, and also $q = q_1$. □

Definition 1.3.1. Let a and b be integers with $a \neq 0$. We say a divides b, denoted by $a \mid b$, if there exists an integer c such that $b = ac$. When a divides b, we say that a is a *divisor* (or *factor*) of b, and b is a *multiple* of a. If a does not divide b, we write $a \nmid b$. If $a \mid b$ and $0 < a < b$, then a is called a *proper divisor* of b. The largest divisor d such that $d \mid a$ and $d \mid b$ is called the *greatest common divisor* (gcd) of a and b. The greatest common divisor of a and b is denoted by $\gcd(a, b)$.

Theorem 1.3.3. Let a, b, q, r be integers with $b > 0$ and $0 \leq r < b$ such that $a = bq + r$. Then $\gcd(a, b) = \gcd(b, r)$.

Proof. Let $X = \gcd(a, b)$ and $Y = \gcd(b, r)$, it suffices to show that $X = Y$. If integer c is a divisor of a and b, it follows from the equation $a = bq + r$ and the divisibility properties that c is a divisor of r also. By the same argument, every common divisor of b and r is a divisor of a. □

Euclid's algorithm is a process by repeatedly applying the division theorem to the pair of integers (a, b), for finding $\gcd(a, b)$.

Definition 1.3.2 (Euclid's algorithm). Let a and b be positive integers. Then the following system of equations for $r_1, r_2, \cdots, r_n$ defines Euclid's algorithm.

$$\left.\begin{array}{lll}
a = bq_0 + r_1 & \Longleftrightarrow & r_1 = a - bq_0 \\
b = r_1q_1 + r_2 & \Longleftrightarrow & r_2 = b - r_1q_1 \\
r_1 = r_2q_2 + r_3 & \Longleftrightarrow & r_3 = r_1 - r_2q_2 \\
r_2 = r_3q_3 + r_4 & \Longleftrightarrow & r_4 = r_2 - r_3q_3 \\
\vdots & \vdots & \vdots \\
r_{n-2} = r_{n-1}q_{n-1} + r_n & \Longleftrightarrow & r_n = r_{n-2} - r_{n-1}q_{n-1} \\
r_{n-1} = r_nq_n + 0 & &
\end{array}\right\} \quad (1.7)$$

This process ends when a remainder of 0 is obtained. This must occur after a finite number of steps; that is, $r_{n+1} = 0$ for some n.

Theorem 1.3.4 (Correctness of Euclid's algorithm). The last nonzero remainder, r_n, in (1.7), is the *greatest common divisor* of a and b. That is,

$$r_n = \gcd(a, b). \tag{1.8}$$

Moreover, values of x and y in

$$\gcd(a, b) = ax + by \tag{1.9}$$

can be obtained by writing each r_i as a linear combination of a and b. (k is a *linear combination* of a and b if $k = ax + by$.)

Proof. Let $d = \gcd(a, b)$. The using the repeated equations in the algorithm, we have:

$$d \mid a \Rightarrow d \mid b \Rightarrow d \mid r_1 \Rightarrow d \mid r_2 \Rightarrow \cdots \Rightarrow d \mid r_n$$

so that $d \leq r_n$. Working backwards, we have:

$$r_n \mid r_{n-1} \Rightarrow r_n \mid r_{n-2} \Rightarrow r_n \mid r_{n-3} \Rightarrow \cdots \Rightarrow r_n \mid a \Rightarrow r_n \mid b.$$

Hence, $r_n \mid a$ and $r_n \mid b$. Since d is the greatest common divisors of a and b, it follows that $r_n \leq d$. Therefore, $r_n = d$. To see that r_n is a linear combination of a and b, we argue by induction that each r_i is a linear combination of a and b. Clearly, r_1 is a linear combination of a and b, since $r_1 = a - bq_0$, so does r_2. In general, r_i is a linear combination of r_{i-1} and r_{i-2}. By the inductive hypothesis we may suppose that these latter two numbers are linear combinations of a and b, and it follows that r_i is also a linear combination of a and b. □

Example 1.3.1. The process of Euclid's algorithm for computing gcd(1281, 243) may be described as follows.

$$\begin{aligned}
1281 &= 243 \cdot 5 + 66 \\
243 &= 66 \cdot 3 + 45 \\
66 &= 45 \cdot 1 + 21 \\
45 &= 21 \cdot 2 + 3 \\
21 &= 3 \cdot 7 + 0
\end{aligned}$$

Thus, $\gcd(1281, 243) = 3$. The process may also be alternatively described as follows.

$$\begin{aligned}
\gcd(1281, 243) &= \gcd(243, 66) && (\text{because } 1281 = 243 \cdot 5 + 66) \\
&= \gcd(66, 45) && (\text{because } 243 = 66 \cdot 3 + 45) \\
&= \gcd(45, 21) && (\text{because } 66 = 45 \cdot 1 + 21) \\
&= \gcd(21, 3) && (\text{because } 45 = 21 \cdot 2 + 3) \\
&= 3 && (\text{because } 21 = 3 \cdot 7 + 0).
\end{aligned}$$

Algorithm 1.3.1 (Euclid's algorithm). Given integers a and b with $a > b > 0$, this algorithm will compute $\gcd(a, b)$.

[1] (Initialization) Set

$$r_{-1} \leftarrow a$$
$$r_0 \leftarrow b,$$
$$i = 0.$$

[2] (Decision) If $r_i = 0$, go to Step [4].

[3] (Main Computation) Set

$$q_i \leftarrow \lfloor r_{i-1}/r_i \rfloor,$$
$$r_{i+1} \leftarrow r_{i-1} - q_i \cdot r_i,$$
$$i \leftarrow i + 1,$$

go to Step [2].

[4] (Exit) Output r_{i-1}.

Example 1.3.2. Let $a = 1281$ and $b = 243$. We tabulate the computing results by executing Algorithm 1.3.1.

q_{i-1}	r_{i-1}
5	243
3	66
1	45
2	21
7	3
	↑
	d

Theorem 1.3.5 (Complexity of Euclid's algorithm). Euclid's algorithm is in $\mathcal{P}$. More specifically, Eucld's algorithm can be performed in $\mathcal{O}((\log_2 a)^3)$ bit operations.

Proof. First of all, if F_{n+1} and F_{n+2} are the successive terms of the Fibonacci sequence with $n > 1$, then the Euclid's algorithm takes exactly n divisions in (1.16) to show $\gcd(F_{n+2}, F_{n+1}) = F_2 = 1$, since

$$\begin{aligned}
F_{n+2} &= F_{n+1} \cdot 1 + F_n, \\
F_{n+1} &= F_n \cdot 1 + F_{n-1}, \\
&\vdots \\
F_4 &= F_3 \cdot 1 + F_2, \\
F_3 &= F_2 \cdot 2 + 0.
\end{aligned}$$

Next, we shall show that the number of divisions needed to find $\gcd(a, b)$ with $a > b > 0$ by Euclid's algorithm is no more than five times the number of digits of b. By replacing r_0 with a and r_1 with b so that every equation in (1.16) has the form

$$r_j = r_{j+1}q_{j+1} + r_{j+2}$$

except the last one which is of the form

$$r_{n-1} = r_n q_n.$$

Note that each of the quotient $q_1, q_2, \cdots, q_{n-1} \geq 1$ and $q_n \geq 2$. Therefore

$$\begin{aligned}
r_n &\geq 1 = F_2 \\
r_{n-1} &\geq 2r_n \geq 2F_2 \geq 2 = F_3 \\
r_{n-2} &\geq r_{n-1} + r_n \geq F_3 + F_2 = F_4 \\
r_{n-3} &\geq r_{n-2} + r_{n-1} \geq F_4 + F_3 = F_5 \\
&\vdots \\
r_2 &\geq r_3 + r_4 \geq F_{n-1} + F_{n-2} = F_n \\
b &= r_1 \geq r_2 + r_3 \geq F_n + F_{n-1} = F_{n+1}.
\end{aligned}$$

Since for all $n \geq 0$,

$$F_n = \frac{\alpha^n - \beta^n}{\sqrt{5}}$$

where

$$\alpha = \frac{1+\sqrt{5}}{2}, \quad \beta = \frac{1-\sqrt{5}}{2}$$

then

$$F_n > \alpha^{n-2}.$$

Thus

$$b > \alpha^{n-2}.$$

Now $\log_{10} b \approx 0.208 > \frac{1}{5}$, so that

$$\log_{10} b > (n-1)\log_{10}\alpha > \frac{n-1}{5}.$$

That is,

$$n - 1 < 5\log_{10} b.$$

Let b have k digits so that $b < 10^k$ and $\log_{10} b < k$. Hence $n - 1 < 5k$. Because k is integer, so $n \leq 5k$. This means that one needs $\mathcal{O}(\log_2 a)$ divisions in Euclid's algorithm to compute $\gcd(a, b)$. As for each division, one needs $\mathcal{O}((\log_2 a)^2)$ bit operations, thus the total bit operations for computing $\gcd(a, b)$ is $\mathcal{O}((\log_2 a)^3)$. Therefore, computing $\gcd(a, b)$ by Euclid's algorithm is in $\mathcal{P}$. □

Euclid's algorithm is found in Book VII, Propositions 1 and 2 of his *Elements*, but it probably wasn't his own invention. Scholars believe that the method was known up to 200 years earlier. However, it first appeared in Euclid's *Elements*, and more importantly, it is the first nontrivial algorithm that has survived to this day. As Hardy remarked in his Apology [137]: Euclid's proof of the infinitude of prime numbers and Pythagoras's proof of the irrationality of $\sqrt{2}$ are as as fresh and significant as when it has discovered – two thousand years have not written a wrinkle on either of them, we could say the same thing for Euclid'd algorithm, as it is indeed as fresh and important as when it has discovered.

It is interesting to note that in almost the same time as Euclid, the ancient Chinese mathematicians also independently discovered an algorithm for computing gcd, called *Mutual Subtraction Algorithm* in Problem 6 of Chapter 1 of the ancient mathematics text *The Nine Chapters on the Mathematical Art* (see [169] and [286]):

> If halving is possible, take half.
>
> Otherwise write down the denominator and the numerator, and subtract the smaller from the greater.
>
> Repeat until both numbers are equal.
>
> Simplify with this common factor.

With this regard, we could call Euclid's algorithm *Mutual Division Algorithm*. To see how similar they are, let us use both methods to compute gcd(195, 135).

$$\begin{array}{ll} 195 = 135 \cdot 1 + 60 & \qquad 195 - 135 = 60 \\ 135 = 60 \cdot 2 + 15 & \qquad 135 - 60 = 75 \\ 60 = 15 \cdot 4 + 0 & \qquad 75 - 60 = 15 \\ & \qquad 60 - 15 = 45 \\ & \qquad 45 - 15 = 30 \\ & \qquad 30 - 15 = 15 \end{array}$$

One may conclude from the above example that Euclid's mutual division algorithm is faster than the Chinese mutual subtraction algorithm. One could also conclude that Euclid's mutual division algorithm is an improvement of the Chinese mutual subtraction algorithm. However, the Chinese mutual subtraction algorithm still has its merits and advantages, which Euclid's mutual division algorithm does not have or is difficult to have; two of the advantages are as follows. Firstly, the Chinese algorithm is easy to use when finding gcd of more than two numbers. This can be shown by the calculation

$$\begin{aligned} \gcd(1008, 1260, 882, 1134) &= \gcd(1008 - 882, 1260 - 1143, 882, 1134 - 882) \\ &= \gcd(126, 126, 882 - 126 \cdot 6, 292 - 126) \\ &= \gcd(126, 126, 126, 126) \\ &= 126 \end{aligned}$$

Secondly and more importantly, the Chinese algorithm can be converted, without any extra afford and labor, to an even more efficient algorithm than Euclid's, now commonly known as binary Euclid's algorithm [49]:

Algorithm 1.3.2 (Binary GCD). Given integers a and b with $a > b > 0$, this algorithm will compute $\gcd(a, b)$ without using divisions and multiplications, except by powers of 2.

[1] Set $t \leftarrow |a - b|$. If $t = 0$, go to Step [4].

[2] While a is even do $a \leftarrow a/2$.

[3] If $a \geq b$ then $a \leftarrow t$ else $b \leftarrow t$, go to Step [1].

[4] Output a (this a is the required $\gcd(a, b)$), and terminate the algorithm.

Remark 1.3.1. As observed by Knuth [169] and Brent [49], or as can be seen from the above discussion, binary Euclid's algorithm is essentially the Chinese Mutual Subtraction Algorithm, which is different from Euclid's algorithm, as Euclid's algorithm does not have the binary feature. This is the *difference* between the ancient west and east algorithms for computing GCD. Readers who are interested in binary GCD algorithm, particularly in the rigorous complexity analysis of the binary GCD algorithm, are advised to consult the excellent paper by Brent [49].

Extended Euclid's Algorithm

Euclid's algorithm can not only be used to compute

$$d = \gcd(a, b),$$

but also be used to solve the linear Diophantine equation of the form

$$d = ax + by,$$

or the linear congruence of the form

$$ax \equiv b \pmod{N}.$$

First let us define the extended Euclid's algorithm.

Definition 1.3.3 (Extended Euclid's algorithm). Let a and b be positive integers with $a \geq b > 0$. Then the following system of equations for $r_1, r_2, \cdots, r_n$ defines extended Euclid's algorithm.

$$
\left.\begin{array}{lcl}
r_1 = a - bq_0 & \Longleftrightarrow & r_1 = a - bq_0 \\
 & & \quad = ax_1 - by_1 \\
r_2 = b - r_1q_1 & \Longleftrightarrow & r_2 = b - r_1q_1 \\
 & & \quad = b - (a - bq_0)q_1 \\
 & & \quad = -aq_1 + (1 + q_0q_1)b \\
 & & \quad = ax_2 + by_2 \\
r_3 = r_1 - r_2q_2 & \Longleftrightarrow & r_3 = r_1 - r_2q_2 \\
 & & \quad = ax_3 + by_3 \\
r_4 = r_2 - r_3q_3 & \Longleftrightarrow & r_4 = r_2 - r_3q_3 \\
 & & \quad = ax_4 + by_4 \\
\vdots & \vdots & \vdots \\
r_n = r_{n-2} - r_{n-1}q_{n-1} & \Longleftrightarrow & r_n = r_{n-2} - r_{n-1}q_{n-1} \\
 & & \quad = ax_n + by_n
\end{array}\right\} \quad (1.10)
$$

Thus, the extended Euclid's algorithm takes input a and b and outputs r_n, x_n and y_n such that

$$r_n = \gcd(a, b) = ax_n + by_n.$$

Example 1.3.3. Find x and y in $\gcd(1281, 243) = 1281x + 243y$. We perform the computation for $\gcd(a, b)$ at the left-hand side and in the same time, the calculation for (x, y) at the right-hand side as follows.

$$
\begin{array}{rcl}
1281 = & 243 \cdot 5 + 66 \Longleftrightarrow & 66 = 1281 - 243 \cdot 5 \\
243 = & 66 \cdot 3 + 45 \Longleftrightarrow & 45 = 243 - 66 \cdot 3 \\
 & & \quad = 243 - 3(1281 - 243 \cdot 5) \\
 & & \quad = 16 \cdot 243 - 3 \cdot 1281 \\
66 = & 45 \cdot 1 + 21 \Longleftrightarrow & 21 = 66 - 45 \cdot 1 \\
 & & \quad = 1281 - 243 \cdot 5 - 16 \cdot 243 + 3 \cdot 1281 \\
 & & \quad = 4 \cdot 1281 - 21 \cdot 243 \\
45 = & 21 \cdot 2 + 3 \Longleftrightarrow & 3 = 45 - 21 \cdot 2 \\
 & & \quad = 16 \cdot 243 - 3 \cdot 1281 - 2(4 \cdot 1281 - 21 \cdot 243) \\
 & & \quad = \underset{\substack{\uparrow \\ x}}{-11} \cdot \underset{\substack{\uparrow \\ a}}{1281} + \underset{\substack{\uparrow \\ y}}{58} \cdot \underset{\substack{\uparrow \\ b}}{243} \\
21 = & 3 \cdot 7 + 0 &
\end{array}
$$

The above process is well suited for computer programming, but of course, if the calculation of (x, y) are by hand, the following (bottom-up) process is convenient.

$$
\begin{aligned}
3 &= 45 - 21 \cdot 2 \\
&= 45 - (66 - 45 \cdot 1) \cdot 2 \\
&= 3 \cdot 45 - 2 \cdot 66 \\
&= 3 \cdot (243 - 66 \cdot 3) - 2 \cdot 66 \\
&= 3 \cdot 243 - 11 \cdot 66 \\
&= 3 \cdot 243 - 11 \cdot (1281 - 243 \cdot 5) \\
&= \underset{\substack{\uparrow \\ x}}{-11} \cdot \underset{\substack{\uparrow \\ a}}{1281} + \underset{\substack{\uparrow \\ y}}{58} \cdot \underset{\substack{\uparrow \\ b}}{243}
\end{aligned}
$$

Algorithm 1.3.3 (Extended Euclid's algorithm). Let $a > b$ be two positive integers, this algorithm finds $d = \gcd(a, b)$ and in the same time, finds x, y in $d = ax + by$.

[1] (Initialization) Let

$$
\begin{array}{ll}
r_0 \leftarrow a & r_1 \leftarrow b, \\
x_0 \leftarrow 1 & y_0 \leftarrow 0, \\
x_1 \leftarrow 0 & y_1 \leftarrow 1, \\
i \leftarrow 1. &
\end{array}
$$

[2] (Decision) If $r_i \bmod r_{i-1} \neq 0$, go to step [4].

[3] (Main Computation) Set

$$
\begin{aligned}
&q_i \leftarrow \lfloor r_{i-1}/r_i \rfloor, \\
&r_{i+1} \leftarrow r_{i-1} - q_i \cdot r_i \\
&x_{i+1} \leftarrow x_{i-1} - q_i \cdot x_i \\
&y_{i+1} \leftarrow y_{i-1} - q_i \cdot y_i \\
&i \leftarrow i + 1
\end{aligned}
$$

[4] (Exit) Output $(q, x, y, d) = (q_{i-1}, x_{i-1}, y_{i-1}, r_{i-1})$.

Example 1.3.4. Let $a = 1281$ and $b = 243$. We tabulate the computing results of (q, d, x, y) at each loop i of the above algorithm 1.3.3:

q	r_i	x_i	y_i
5	243	0	1
3	66	1	-5
1	45	-3	16
2	21	4	-21
7	3	-11	58
	↑	↑	↑
	d	x	y

That is,

$$\gcd(a, b) = ax + by \longleftrightarrow 3 = 1281(-11) + 243 \cdot 58.$$

The following example shows how to use extended Euclid's algorithm to solve the linear congruence of the form $ax \equiv b \pmod{N}$.

Example 1.3.5. Find the x in $5033464705x \equiv 1 \pmod{12347}$. First notice that

$$5033464705x \equiv 1 \pmod{12347} \Longleftrightarrow 5033464705x - 12347y = 1.$$

Of course here we are only interested in the value of x, not y. We tabulate the results of each computation step in Algorithm 1.3.3 as follows.

q	x_1	y_1	r_1
-	1	0	5033464705
407667	0	1	12347
48	1	-407667	256
4	-48	19568017	59
2	193	-78679735	20
1	-434	176927487	19
19	627	-255607222	1
	$\uparrow$		$\uparrow$
	x		d

Thus, $x = 627$ is a solution to the congruence equation $5033464705x \equiv 1 \pmod{12345}$. This can be verified quickly to be true, since $5033464705 \cdot 627 \equiv 1 \pmod{12345}$.

How about if $\gcd(a, N) > 1$ when solving $ax \equiv b \pmod{N}$? For example, find the x in $5033464705x \equiv 7 \pmod{12345}$, where $\gcd(5033464705, 12345) = 5 > 1$. There are two cases:

(1) If $d \nmid b$, then there is no solution to the congruence.

(2) If $d \mid b$, then we need to consider whether or not there is a solution to the new congruence $(a/d)x \equiv b/d \pmod{N/d}$.

Thus, for $5033464705x \equiv 7 \pmod{12345}$, since $\gcd(5033464705, 12345) = 5$, we get a new congruence $(5033464705/5)x \equiv 7/5 \pmod{12345/5}$, which is $164x \equiv 989 \pmod{2469}$.

Theorem 1.3.6. Let $d = \gcd(a, N)$.

(1) If $d \nmid b$, then the linear congruence

$$ax \equiv b \pmod{N} \tag{1.11}$$

has no solution.

(2) The linear congruence $ax \equiv b \pmod{N}$ has solutions if and only if $d \mid b$.

(3) The linear congruence $ax \equiv b \pmod{N}$ has exactly one solution if $\gcd(a, N) = 1$.

(4) If $d \mid b$, then the linear congruence $ax \equiv b \pmod{N}$ has exactly d solutions modulo N. These are given by

$$t,\ t+\frac{N}{d},\ t+\frac{2N}{d},\ \cdots,\ t+\frac{(d-1)N}{d} \tag{1.12}$$

where t is the solution, unique modulo N/d, of the linear congruence

$$\frac{a}{d}x \equiv \frac{b}{d} \left(\bmod \frac{N}{d}\right). \tag{1.13}$$

Example 1.3.6. Solve $803x \equiv 22 \pmod{154}$. Notice first that

$$803x \equiv 22 \pmod{154} \Longleftrightarrow 803x - 154y = 22.$$

Use the extended Euclid's algorithm 1.3.3, we get:

q	u_1	u_2	u_3
-	1	0	803
5	0	1	154
4	1	-5	33
1	-4	21	22
2	5	-26	11
	↑		↑
	x		d

Since $d \mid b$, and $d = 11$ and $x = 5$, the congruence has exactly 11 solutions $x + iN/d$ for $i = 0, 1, 2, \cdots, d-1$, namely,

$$x = \{5, 19, 33, 47, 61, 75, 89, 103, 117, 131, 145\}.$$

Theorem 1.3.7. The complexity of extended Euclid's algorithm is in $\mathcal{P}$.

Continued Fraction Algorithm

Euclid's algorithm is also intimately connected with the continued fraction algorithm, and in fact, when performing the extended Euclid's algorithm 1.3.3 on a, b, we get a sequence of numbers $q_0, q_1, \cdots, q_n$, which form a continued fraction of a/b as follows:

$$a/b = q_0 + \cfrac{1}{q_1 + \cfrac{1}{q_2 + \cfrac{1}{\ddots\, q_{n-1} + \cfrac{1}{q_n}}}} \tag{1.14}$$

Let us look at Example 1.3.6 again. If we only take the values of the q column, ignoring all other values in other columns, we get:

$$\frac{803}{154} = 5 + \cfrac{1}{4 + \cfrac{1}{1 + \cfrac{1}{2}}} \tag{1.15}$$

Of course, if we are only interested in the values of q for the rational number a/b, we can just use Algorithm 1.3.1 or Algorithm 1.3.3 to generate the continued fraction for a/b. In what follows, we consider how to express a general real number x, e.g., $\sqrt{3}$, rather than just a rational number a/b, as a continued fraction. Remarkably enough, the continued fraction algorithm for real numbers is still the same as Euclid's algorithm, except for notation.

Definition 1.3.4 (Continued fraction algorithm). Let x be a real number. Let $q_0 = \lfloor x \rfloor$. Then the continued fraction algorithm is defined as follows.

$$\left.\begin{array}{ll} x = q_0 + \xi_0, & 0 < \xi_0 < 1 \\ \dfrac{1}{\xi_0} = x_1 = q_1 + \xi_1, & 0 < \xi_1 < 1 \\ \dfrac{1}{\xi_1} = x_2 = q_2 + \xi_2, & 0 < \xi_2 < 1 \\ \dfrac{1}{\xi_2} = x_3 = q_3 + \xi_3, & 0 < \xi_3 < 1 \\ \vdots & \vdots \end{array}\right\} \tag{1.16}$$

The algorithm continues as long as $\xi_n \neq 0$. Whenever $\xi_n = 0$ for any positive integer n, the algorithm terminates and

$$x = q_0 + \cfrac{1}{q_1 + \cfrac{1}{q_2 + \cfrac{1}{\ddots\, q_{n-1} + \cfrac{1}{q_n}}}} \tag{1.17}$$

where $q_0, q_1, \cdots, q_{n-1}, q_n$ are taken directly from Euclid's algorithm expressed in (1.16), and are called the *partial quotients* of the continued fraction, or briefly as,

$$x = q_0 + \frac{1}{q_1+}\frac{1}{q_2+}\cdots\frac{1}{q_{n-1}+}\frac{1}{q_n} \tag{1.18}$$

or even more briefly as

$$x = [q_0, q_1, q_2, \cdots q_{n-1}, q_n]. \tag{1.19}$$

Just the same as q_n are the partial quotients of the continued fraction of x, the numbers x_n are the *complete quotients* of the continued fraction of x. If each q_i is an integer, then the continued fraction is called *simple*; a simple continued fraction can either be *finite* or *infinite*.

Theorem 1.3.8. Every rational number can be represented by a finite simple continued fraction in exactly two ways, one with an odd number of terms and one with even number of terms. Every irrational number can be expressed in exactly one way as an infinite simple continued fraction.

For example, let $x = a/b$ be a rational number, then Euclid's algorithm for for computing $\gcd(a, b)$ will produce the continued fraction expansion of a/b in the same time as follows:

$$\frac{a}{b} = q_0 + \cfrac{1}{q_1 + \cfrac{1}{q_2 + \cfrac{1}{\ddots\, q_{n-1} + \cfrac{1}{q_n}}}} \tag{1.20}$$

where $q_0, q_1, \cdots, q_{n-1}, q_n$ are taken directly from Euclid's algorithm expressed in (1.16), or in (1.14).

The continued fraction algorithm defined in (1.16) can be conveniently modified to be as follows:

$$\left.\begin{array}{ll} q_0 = \lfloor x_0 \rfloor, & x_1 = \dfrac{1}{x_0 - q_0} \\ q_1 = \lfloor x_1 \rfloor, & x_2 = \dfrac{1}{x_1 - q_1} \\ \vdots & \vdots \\ q_n = \lfloor x_n \rfloor, & x_{n+1} = \dfrac{1}{x_n - q_n} \\ q_{n+1} = \lfloor x_{n+1} \rfloor, & x_{n+2} = \dfrac{1}{x_{n+1} - q_{n+1}} \\ \vdots & \vdots \end{array}\right\} \tag{1.21}$$

which is, in turn, can be easily implemented as follows.

Algorithm 1.3.4 (Continued fraction algorithm). Given a real number x, this algorithm will compute and output the partial quotients $q_0, q_1, q_2, \cdots, q_n$ of the continued fraction x.

[1] (Initialization) Set

$i \leftarrow 0$,
$x_i \leftarrow x$,
$q_i \leftarrow \lfloor x_i \rfloor$,
print(q_i).

[2] (Decision) If $x_i = q_i$ go to Step [4].

[3] (Main Computation) Set

$$x_{i+1} \leftarrow \frac{1}{x_i - q_i},$$
$$i \leftarrow i+1,$$
$$q_i \leftarrow \lfloor x_i \rfloor,$$
print(q_i),
go to Step [2].

[4] Exit.

Example 1.3.7. Let $x = 160523347/60728973$. Then by applying Algorithm 1.3.4, we get $160523347/60728973 = [2, 1, 1, 1, 4, 12, 102, 1, 1, 2, 3, 2, 2, 36]$. That is,

$$\frac{160523347}{60728973} = 2 + \cfrac{1}{1 + \cfrac{1}{1 + \cfrac{1}{1 + \cfrac{1}{4 + \cfrac{1}{12 + \cfrac{1}{102 + \cfrac{1}{1 + \cfrac{1}{1 + \cfrac{1}{2 + \cfrac{1}{3 + \cfrac{1}{2 + \cfrac{1}{2 + \cfrac{1}{36}}}}}}}}}}}}}$$

Theorem 1.3.9. The continued fraction algorithm is in $\mathcal{P}$.

Definition 1.3.5. If each q_i is an integer, the continued fraction is called *simple*; a simple continued fraction can either be *finite* or *infinite*. A continued fraction formed from $[q_0, q_1, q_2, \cdots q_{n-1}, q_n]$ by neglecting all of the terms after a given term is called a *convergent* of the original continued fraction. If we denote the k-th convergent by $C_k = \dfrac{P_k}{Q_k}$, then

$$(1) \quad \begin{cases} C_0 = \dfrac{P_0}{Q_0} = \dfrac{q_0}{1}; \\ C_1 = \dfrac{P_1}{Q_1} = \dfrac{q_0 q_1 + 1}{q_1}; \\ \vdots \\ C_k = \dfrac{P_k}{Q_k} = \dfrac{q_k P_{k-1} + P_{k-2}}{q_k Q_{k-1} + Q_{k-2}}, \quad \text{for } k \geq 2. \end{cases}$$

(2) If $P_k = q_k Q_{k-1} + Q_{k-2}$ and $Q_k = q_k P_{k-1} + P_{k-2}$, then $\gcd(P_k, Q_k) = 1$.

(3) $P_k Q_{k-1} - P_{k-1} Q_k = (-1)^{k-1}$, for $k \geq 1$.

If $\gcd(a, b) = d$, then the convergent to a/b can be used to solve the linear Diophantine equations $ax + by = d$.

Theorem 1.3.10. Let the convergents of the finite continued fraction of a/b be as follows:

$$\left[\frac{P_0}{Q_0}, \frac{P_1}{Q_1}, \cdots, \frac{P_{n-1}}{Q_{n-1}}, \frac{P_n}{Q_n}\right] = \frac{a}{b}. \tag{1.22}$$

Then the integer solution in x and y of the equation $ax - by = d$ is

$$\left.\begin{aligned} x &= (-1)^{n-1} Q_{n-1} \\ y &= (-1)^{n-1} P_{n-1} \end{aligned}\right\} \tag{1.23}$$

Corollary 1.3.1. The integer solution in x and y for equations $ax + by = d$, $-ax - by = d$ and $-ax + by = d$ are as follows:

$$\left.\begin{aligned} x &= (-1)^{n-1} q_{n-1} \\ y &= (-1)^{n} p_{n-1} \end{aligned}\right\} \tag{1.24}$$

for $ax + by = d$.

$$\left.\begin{aligned} x &= (-1)^{n} q_{n-1} \\ y &= (-1)^{n-1} p_{n-1} \end{aligned}\right\} \tag{1.25}$$

for $-ax - by = d$.

$$\left.\begin{aligned} x &= (-1)^{n} q_{n-1} \\ y &= (-1)^{n} p_{n-1} \end{aligned}\right\} \tag{1.26}$$

for $-ax + by = d$.

Remark 1.3.2. We have already seen how to use extended Euclid's algorithm to solve the linear Diophantine equations $ax + by = d$. The continued fraction algorithm just discussed turns out to be equivalent to extended Euclid's algorithm, since the continued fraction of a/b is derived from Euclid's algorithm. They are all efficient and can be done in polynomial-time.

Example 1.3.8. Use the continued fraction method to solve the following linear Diophantine equation:

$$364x - 227y = 1.$$

Since $364/227$ can be expanded as a finite continued fraction with convergents

$$\left[1,\ 2,\ \frac{3}{2},\ \frac{5}{3},\ \frac{8}{5},\ \frac{85}{53},\ \frac{93}{58},\ \frac{364}{227}\right]$$

we have

$$x = (-1)^{n-1} q_{n-1} = (-1)^{7-1} \cdot 58 = 58,$$
$$y = (-1)^{n-1} p_{n-1} = (-1)^{7-1} \cdot 93 = 93.$$

That is,

$$364 \cdot 58 - 227 \cdot 93 = 1.$$

Example 1.3.9. Use the continued fraction method to solve the following linear Diophantine equation:

$$20719x + 13871y = 1.$$

Since 20719/13871 can be expanded as a finite simple continued fraction with convergents

$$\left[1, \frac{3}{2}, \frac{118}{79}, \frac{829}{555}, \frac{947}{634}, \frac{1776}{1189}, \frac{2723}{1823}, \frac{4499}{3012}, \frac{20719}{13871}\right],$$

we have

$$x = (-1)^{n-1} q_{n-1} = (-1)^{8-1} \cdot 3012 = -3012,$$
$$y = (-1)^{n} p_{n-1} = (-1)^{8} \cdot 4499 = 4499.$$

That is,

$$20719 \cdot (-3012) + 13871 \cdot 4499 = 1.$$

The Chinese Remainder Theorem (CRT)

We now consider the solutions to systems of linear congruences. In around 200BC, the Chinese mathematician Sun Tsu proposed a problem in his classic three-volume mathematics book *Mathematical Manual*: find a number that leaves a remainder of 2 when divided by 3, a remainder of 3 when divided by 5, and a remainder of 2 when divided by 7. In modern algebraic language, it is to find the smallest positive integer satisfying the following systems of congruences:

$$x \equiv 2 \pmod{3},$$
$$x \equiv 3 \pmod{5},$$
$$x \equiv 2 \pmod{7}.$$

The general form of the problem is as follows.

Theorem 1.3.11 (The Chinese Remainder Theorem (CRT)). If m_1, $m_2, \cdots, m_n$ are pairwise relatively prime and greater than 1, and $a_1, a_2, \cdots$, a_n are any integers, then there is a solution x to the following simultaneous congruences:

$$\left.\begin{array}{l} x \equiv a_1 \pmod{m_1}, \\ x \equiv a_2 \pmod{m_2}, \\ \vdots \\ x \equiv a_n \pmod{m_n}. \end{array}\right\} \quad (1.27)$$

If x and x' are two solutions, then $x \equiv x' \pmod{M}$, where $M = m_1 m_2 \cdots m_n$.

Proof. Existence: Let us first solve a special case of the simultaneous congruences (1.27), where i is some fixed subscript,

$$a_i = 1, \ a_1 = a_2 = \cdots = a_{i-1} = a_{i+1} = \cdots = a_n = 0.$$

Let $k_i = m_1 m_2 \cdots m_{i-1} m_{i+1} \cdots m_n$. Then k_i and m_i are relatively prime, so we can find integers r and s such that $rk_i + sm_i = 1$. This gives the congruences:

$$\begin{array}{rcl} rk_i & \equiv & 0 \pmod{k_i}, \\ rk_i & \equiv & 1 \pmod{m_i}. \end{array}$$

Since $m_1, m_2, \cdots, m_{i-1}, m_{i+1}, \cdots m_n$ all divide k_i, it follows that $x_i = rk_i$ satisfies the simultaneous congruences:

$$\begin{array}{rcl} x_i & \equiv & 0 \pmod{m_1}, \\ x_i & \equiv & 0 \pmod{m_2}, \\ & \vdots & \\ x_i & \equiv & 0 \pmod{m_{i-1}}, \\ x_i & \equiv & 1 \pmod{m_i}, \\ x_i & \equiv & 0 \pmod{m_{i+1}}, \\ & \vdots & \\ x_i & \equiv & 0 \pmod{m_n}. \end{array}$$

For each subscript i, $1 \le i \le n$, we find such an x_i. Now, to solve the system of the simultaneous congruences (1.27), set $x = a_1 x_1 + a_2 x_2 + \cdots + a_n x_n$. Then $x \equiv a_i x_i \equiv a_i \pmod{m_i}$ for each i, $1 \le i \le n$, therefore x is a solution of the simultaneous congruences.

Uniqueness: Let x' be another solution to the simultaneous congruences (1.27), but different from the solution x, so that $x' \equiv x \pmod{m_i}$ for each x_i. Then, $x - x' \equiv 0 \pmod{m_i}$ for each i. So, m_i divides $x - x'$ for each i; hence the least common multiple of all the m_j's divides $x - x'$. But since the m_i are pairwise relatively prime, this least common multiple is the product M. So, $x \equiv x' \pmod{M}$. □

Corollary 1.3.2. If the system of the linear simultaneous congruences (1.27) is soluble, then its solution can be conveniently written as follows:

$$x \equiv \sum_{i=1}^{n} a_i M_i M_i' \pmod{m} \tag{1.28}$$

where

$$\begin{aligned} m &= m_1 m_2 \cdots m_n, \\ M_i &= m/m_i, \\ M_i' &= M_i^{-1} \pmod{m_i}, \end{aligned}$$

for $i = 1, 2, \cdots, n$.

Theorem 1.3.12. Chinese Remainder Theorem is in $\mathcal{P}$.

Example 1.3.10. Use the Chinese Remainder Theorem to solve

$$\begin{cases} x \equiv 12 \pmod{101}, \\ x \equiv 93 \pmod{235}, \\ x \equiv 28 \pmod{719}. \end{cases}$$

As $101, 235, 719$ are pairwise relatively prime, we use Equation (1.28) to find the solution x:

$$\begin{aligned} x &\equiv \sum_{i=1}^{3} a_i M_i M_i' \pmod{m} \\ &\equiv 12 \cdot M_1 \cdot M_1' + 93 \cdot M_2 \cdot M_2' + 28 \cdot M_3 \cdot M_3' \pmod{M_1 M_2 M_3} \\ &\equiv 12 \cdot 168965 \cdot 63 + 93 \cdot 72619 \cdot 59 + 28 \cdot 23735 \cdot 90 \pmod{17065465} \\ &= 5784383. \end{aligned}$$

Fast Modular Exponentiation

The basic idea for fast exponentiation $x^e \pmod{N}$ is as follows: First, e is written in its binary form $e = (e_\beta e_{\beta-1} \cdots e_1 e_0)_2$, which is then stored in a memory. Then, according to the binary bits $e_\beta, e_{\beta-1}, \cdots, e_1, e_0$ (starting from e_β to e_0, i.e., from the most significant bit to the least significant bit), the related operations are carried out. For example, let $e = 100 = (1100100)_2$, then the related squaring and multiplication operations will be carried out as follows:

e_6	1	x	x
e_5	1	$x^2 \cdot x$	x^3
e_4	0	$(x^2 \cdot x)^2$	x^6
e_3	0	$((x^2 \cdot x)^2)^2$	x^{12}
e_2	1	$(((x^2 \cdot x)^2)^2)^2 \cdot x$	x^{25}
e_1	0	$((((x^2 \cdot x)^2)^2)^2 \cdot x)^2$	x^{50}
e_0	0	$(((((x^2 \cdot x)^2)^2)^2 \cdot x)^2)^2$	x^{100}

$$(((((x^2 \cdot x)^2)^2)^2 \cdot x)^2)^2 = x^{100}$$

The general algorithm for computing $y \equiv x^e \pmod N$ may be given as follows:

Algorithm 1.3.5 (Fast modular exponentiation for $y \equiv x^e \pmod n$). This algorithm will find the modular exponentiation

$$y \equiv x^e \pmod N$$

where $x, e, N \in \mathbb{N}$ with $N > 1$.

[1] (Precomputation) Let

$$e = e_\beta e_{\beta-1} \cdots e_1 e_0 \tag{1.29}$$

be the binary representation of e (i.e., e has $\beta + 1$ bits). For example, for $562 = (1000110010)_2$, we have $\beta = 9$ and

$$\begin{array}{cccccccccc} 1 & 0 & 0 & 0 & 1 & 1 & 0 & 0 & 1 & 0 \\ \uparrow & \uparrow & \uparrow & \uparrow & \uparrow & \uparrow & \uparrow & \uparrow & \uparrow & \uparrow \\ e_9 & e_8 & e_7 & e_6 & e_5 & e_4 & e_3 & e_2 & e_1 & e_0 \end{array}$$

[2] (Initialization) Set $y \leftarrow 1$.

[3] (Squaring and Multiplication) Compute $y \equiv x^e \pmod N$ in the following way:

```
for i from β down to 0 do
    y ← y^2 mod n (squaring)
    if e_i = 1 then
            y ← y · x mod N (multiplication)
```

[4] (Exit) Print y and terminate the algorithm.

Theorem 1.3.13. Let x, e and N be positive integers with $N > 1$. Then the modular exponentiation $x^e \bmod N$ can be computed in $\mathcal{O}(\log e)$ arithmetic operations and $\mathcal{O}\left((\log e)(\log N)^2\right)$ bit operations. That is, modular exponentiation is in $\mathcal{P}$.

Proof. We first find the least positive residues of $x, x^2, x^4, \cdots, x^{2^k}$ modulo n, where $2^k \leq e < 2^{k+1}$, by successively squaring and reducing modulo n. This requires a total of $\mathcal{O}\left((\log e)(\log n)^2\right)$ bit operations, since we perform $\mathcal{O}(\log e)$

squarings modulo N, each requiring $\mathcal{O}((\log N)^2)$ bit operations. Next, we multiply together the least positive residues of the integers x^{2^i} corresponding to the binary bits of e which are equal to 1, and reduce modulo n. This also requires $\mathcal{O}\left((\log e)(\log N)^2\right)$ bit operations, since there are at most $\mathcal{O}(\log e)$ multiplications, each requiring $\mathcal{O}((\log n)^2)$ bit operations. Therefore, a total of $\mathcal{O}\left((\log e)(\log N)^2\right)$ bit operations are needed to find the least positive residue of $x^e \bmod N$. □

Fast Group Operations on Elliptic Curves

The idea of fast exponentiation $x^e \pmod N$ can be easily extended to the group operation (point addition) $Q \equiv kP \pmod N$ where P and Q are points on elliptic curve $E : y^2 \equiv x^3+ax+b \pmod N$. Suppose E is an elliptic curve as shown in Figure 1.12. A straight line (non-vertical) L connecting points P and Q intersects the elliptic curve E at a third point R, and the point $P \oplus Q$ is the reflection of R in the X-axis. That is, if $R = (x_3, y_3)$, then $P \oplus Q = (x_3, -y_3)$ is the reflection of R in the X-axis. Note that a vertical line, such as L' or L'', meets the curve at two points (not necessarily distinct), and also at the point at infinity $\mathcal{O}_E$ (we may think of the point at infinity as lying far off in the direction of the Y-axis). The line at infinity meets the curve at the point $\mathcal{O}_E$ three times. Of course, the non-vertical line meets the curve in three points in the XY plane. Thus, every line meets the curve in three points.

The algebraic formula for computing $P_3(x_3, y_3) = P_1(x_1, y_1) + P_2(x_2, y_2)$ on E is as follows:

$$(x_3, y_3) = (\lambda^2 - x_1 - x_2,\ \lambda(x_1 - x_3) - y_1),$$

where

$$\lambda = \begin{cases} \dfrac{3x_1^2 + a}{2y_1} & \text{if } P_1 = P_2 \\[2ex] \dfrac{y_2 - y_1}{x_2 - x_1} & \text{otherwise.} \end{cases}$$

The idea for fast computing $Q = kP$ over an elliptic curve E is similar to that of computing $y = x^k$ over $\mathbb{N}$. For example, to compute $Q = 105P$, we first let

$$k = 105 = (1101001)_2$$

and then perform the operations as follows starting from e_6 to e_0:

$$\begin{array}{llll}
1: & Q \leftarrow P + 2Q & \Rightarrow Q \leftarrow P & \Rightarrow Q = P \\
1: & Q \leftarrow P + 2Q & \Rightarrow c \leftarrow P + 2P & \Rightarrow Q = 3P \\
0: & Q \leftarrow 2Q & \Rightarrow Q \leftarrow 2(P + 2P) & \Rightarrow Q = 6P \\
1: & Q \leftarrow P + 2Q & \Rightarrow Q \leftarrow P + 2(2(P + 2P)) & \Rightarrow Q = 13P \\
0: & Q \leftarrow 2Q & \Rightarrow Q \leftarrow 2(P + 2(2(P + 2P))) & \Rightarrow Q = 26P \\
0: & Q \leftarrow 2Q & \Rightarrow Q \leftarrow 2(2(P + 2(2(P + 2P)))) & \Rightarrow Q = 52P \\
1: & Q \leftarrow P + 2Q & \Rightarrow Q \leftarrow P + 2(2(2(P + 2(2(P + 2P))))) & \Rightarrow Q = 105P.
\end{array}$$

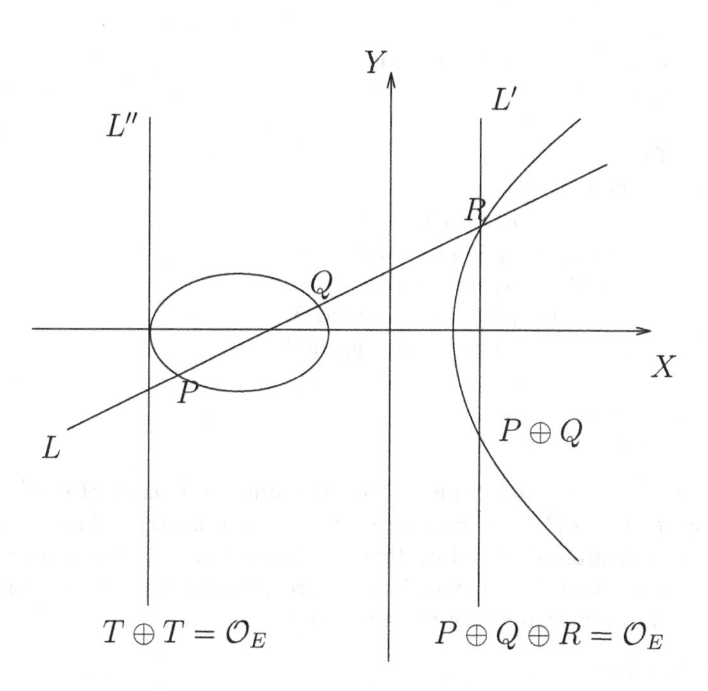

Figure 1.12. Geometric composition laws of an elliptic curve

This gives the required result $Q = P + 2(2(2(P + 2(2(P + 2P))))) = 105P$. The above is just the idea of the fast group operations on E, as we have not considered P and Q as the actual points (i.e., the values of the x and y co-ordinates) on E. The following algorithm, however, provides a full version of the fast group operation $Q \equiv kP \pmod{N}$ on E

Algorithm 1.3.6 (Fast group operations kP on elliptic curves). This algorithm will compute the point $kP \bmod N$, where $k \in \mathbb{Z}^+$ and P is an initial point (x, y) on an elliptic curve $E: y^2 = x^3 + ax + b$ over $\mathbb{Z}/N\mathbb{Z}$; if we require E over $\mathbb{Q}$, just compute kP, rather than $kP \bmod N$. Let the initial point $P = (x_1, y_1)$, and the result point $P = (x_c, y_c)$.

[1] (Precomputation) Write k in the following binary expansion form $k = e_{\beta-1}e_{\beta-2}\cdots e_1e_0$. (Suppose k has β bits).

[2] [Initialization] Initialize the values for a, x_1 and y_1. Let $(x_c, y_c) = (x_1, y_1)$; this is exactly the computation task for e_1 (e_1 always equals 1).

[3] (Doublings and Additions) Computing $kP \bmod N$:

for i from $\beta - 2$ down to 0 do
  $m_1 \leftarrow 3x_c^2 + a \bmod N$
  $m_2 \leftarrow 2y_c \bmod N$
  $M \leftarrow m_1/m_2 \bmod N$
  $x_3 \leftarrow M^2 - 2x_c \bmod N$
  $y_3 \leftarrow M(x_c - x_3) - y_c \bmod N$
  $x_c \leftarrow x_3$
  $y_c \leftarrow y_3$
  if $e_i = 1$
    then $c \leftarrow 2c + P$
      $m_1 \leftarrow y_c - y_1 \bmod N$
      $m_2 \leftarrow x_c - x_1 \bmod N$
      $M \leftarrow m_1/m_2 \bmod N$
      $x_3 \leftarrow M^2 - x_1 - x_c \bmod N$
      $y_3 \leftarrow M(x_1 - x_3) - y_1 \bmod N$
      $x_c \leftarrow x_3$
      $y_c \leftarrow y_3$
    else $c \leftarrow 2c$

[4] (Exit) Print (c, x_c, y_c) and terminate the algorithm. The value of c will be the value of $Q = kP$, and the values of x_c, y_c will be the values of the x and y coordinates of Q. (Note that this algorithm will stop whenever $m_1/m_2 \equiv \mathcal{O}_E \pmod N$, that is, it will stop whenever a modular inverse does not exit at any step of the computation.)

Exercise 1.3.1. Let

$$E: \; y^2 = x^3 - x - 1$$

be an elliptic curve over $\mathbb{Z}/1098413\mathbb{Z}$ and $P = (0, 1)$ a point on E. Use Algorithm 1.3.6 to compute the coordinates (x, y) of the points kP on E over $\mathbb{Z}/1098413\mathbb{Z}$ for $k = 8, 31, 92, 261, 513, 875, 7892, 10319$. Find also the smallest integral values of k such that $kP = (467314, 689129)$ and $kP = (965302, 895958)$, respectively.

Theorem 1.3.14. Suppose that an elliptic curve E is defined over a finite field $\mathbb{Z}_n$. Given $P \in E$, the coordinates of kP can be computed by Algorithm 1.3.6 in $\mathcal{O}(\log k)$ group operations and $\mathcal{O}\left((\log k)(\log N)^3\right)$ bit operations. That is, computing kP over E is in $\mathcal{P}$.

Primality Testing

The Primality Testing Problem (PTP) may be described as follows:

$$\begin{cases} \text{Input}: & N \in \mathbb{Z}_{>1} \\ \text{Output}: & \begin{cases} \text{Yes}, & \text{if } N \in \text{Primes} \\ \text{No}, & \text{otherwise.} \end{cases} \end{cases} \tag{1.30}$$

As the RSA cryptosystem is based on the use of large prime numbers, primality testing is a very important task in the design of RSA. There are many efficient (polynomial-time) algorithms, deterministic and randomized, for primality testing; three of them will be briefly introduced here.

Theorem 1.3.15 (Prime Number Theorem). Let $\pi(x)$ be the number of primes up to x. Then

$$\pi(x) \sim x/\ln x. \tag{1.31}$$

That is,

$$\lim_{x\to\infty} \frac{\pi(x)}{x/\ln x} = 1. \tag{1.32}$$

Remark 1.3.3. The Prime Number Theorem (PNT) was postulated by Gauss (1777–1855) in 1792 on numerical evidence. It is known that Gauss constructed by hand a table of all primes up to three million, and investigated the number of primes occurring in each group of 1000. It was also conjectured by Legendre (1752–1833) before Gauss, in a different form, but of course, both Legendre and Gauss were unable to prove PNT. It was in 1896 that the French mathematician Hadamard (1865–1963) and the Belgian mathematician de la Vallée-Poussin (1866–1962) independently proved the theorem, using some advanced techniques from complex analysis; an elementary proof (i.e., without using calculus) was given independently by Selberg (1917–2007) and Erdös (1913–1996) in 1949. There are thousands of theorems about prime numbers, only this theorem is called the *Prime Number Theorem.*

According to the prime number theorem, the probability of a randomly chosen number N to be prime is about $1/\ln N$. In August 2002 Agrawal, Kayal and Saxena [7] proposed a deterministic polynomial-time test (AKS test for short) for primality, relying on no unproved assumptions. It is not a great surprise that such a test exists, since Dixon [104] predicated in 1984 that "the prospect for a polynomial-time algorithm for proving primality seems fairly good, but it may turn out that, on the contrary, factoring is NP-hard". However, the simplicity of the test is indeed a big surprise.

Algorithm 1.3.7 (AKS test). Give a positive integer $N > 1$, this algorithm will decide whether or not N is prime in deterministic polynomial-time.

[1] If $N = a^b$ with $a \in \mathbb{N}$ and $b > 1$, then output COMPOSITE.

[2] Find the smallest r such that $\mathrm{ord}_r(N) > 4(\log N)^2$.

[3] If $1 < \gcd(a, N) < N$ for some $a \le r$, then output COMPOSITE.

[4] If $N \le r$, then output PRIME.

[5] For $a = 1$ to $\lfloor 2\sqrt{\phi(r)} \log N \rfloor$ do
if $(x-a)^N \not\equiv (x^N - a) \pmod{x^r - 1, N}$,
then output COMPOSITE.

[6] Output PRIME.

The test is indeed very simple and short (with only 6 statements), possibly the shortest algorithm for a (big and previously) unsolved problem ever!

Theorem 1.3.16. The AKS test returns PRIME if and only if n is prime, and runs in $\mathcal{P}$ (more specifically in $\mathcal{O}((\log N)^{10.5})$; the original AKS test runs in $\mathcal{O}((\log N)^{12})$.

Proof. See [7] or [337]. □

Clearly, the AKS test is a rather high order polynomial-time algorithm, which is obviously not so practically useful. In practice, we would like to have a combined use of a Rabin-Miller test (in $\mathcal{RP}$) and an Atkin-Morain ECPP test (in $\mathcal{ZPP}$), which should be almost infallible for all practical purposes. Thus, in practice, we can first perform a Rabin-Miller test on N (Maple provides an even better primality test `isprime`, which is in fact a combined test of a Rabin-Miller test and a Lucas test). If N can pass the Rabin-Miller test or the Maple `isprime` test, then we perform one more test, namely, the ECPP test. If N can still pass the ECPP test, then it is a prime for sure, as ECPP is a $\mathcal{ZPP}$ test.

Algorithm 1.3.8 (Miller-Rabin test). This algorithm will test N for primality with high probability:

[1] Let N be an odd number, and the base b a random number in the range $1 < b < N$. Find j and d with d odd, so that $N - 1 = 2^j d$.

[2] Set $i \leftarrow 0$ and $y \leftarrow b^d \pmod{N}$.

[3] If $i = 0$ and $y = 1$, or $y = N - 1$, then terminate the algorithm and output "n is probably prime". If $i > 0$ and $y = 1$ goto [5].

[4] $i \leftarrow i + 1$. If $i < j$, set $y \equiv y^2 \pmod{N}$ and return to [3].

[5] Terminate the algorithm and output "N is definitely not prime".

Theorem 1.3.17. The Miller-Rabin test runs in expected time $\mathcal{O}((\log N)^3)$, and if the generalized Riemann hypothesis is true then the test can be run deterministically in time $\mathcal{O}((\log N)^5)$.

Proof. Let $b < N$ be a positive integer. Then the primality of N can be determined in expected time $\mathcal{O}((\log N)^3)$, since only $\mathcal{O}(\log N)$ modular exponentiations may be needed and each of them takes $\mathcal{O}((\log N)^2)$ bit operations, resulting $\mathcal{O}(\log N) \cdot \mathcal{O}((\log N)^2) = \mathcal{O}((\log N)^3)$. If the generalized Riemann hypothesis (for L-functions of quadratic characters [335]) is true and N is composite, then there is a b with $1 < b < 2(\log N)^2$ [267] such that N fails the Miller-Rabin test for b. To discover such a b requires $\mathcal{O}((\log N)^3) \cdot \mathcal{O}((\log N)^2) = \mathcal{O}((\log N)^5)$ bit operations. □

The basic theory of elliptic curve tests may be given by the following theorem:

Theorem 1.3.18 (Cox [89]). Let $N \in \mathbb{N}_{>13}$ with $\gcd(N, 6) = 1$, and let $E : y^2 \equiv x^3 + ax + b \pmod N$ be an elliptic curve over $\mathbb{Z}_N$. Suppose that

1) $N + 1 - 2\sqrt{N} \le |E(\mathbb{Z}_N)| \le N + 1 + 2\sqrt{N}$.
2) $|E(\mathbb{Z}_N)| = 2q$, with q an odd prime.

If $P \ne \mathcal{O}_E$ is a point on E and $qP = \mathcal{O}_E$ on E, then N is prime.

There are two main types of elliptic curve tests: Goldwasser-Kilian [123] and Atkin-Morain [14]. The currently fastest version of Atkin-Morain test, fastECPP [222], runs in expected time $\mathcal{O}((\log N)^4)$. So, for all practical purposes, we could just simply use a combined test of a probabilistic such as Miller-Rabin or Maple `isprime`) and an elliptic curve test such as ECPP as follows.

Algorithm 1.3.9 (Practical primality testing). Given a random odd positive integer n, this algorithm will make a combined use of probabilistic tests and elliptic curve tests to determine whether or not n is prime:

[1] (Primality Testing – Probabilistic Testing) Use a combination of the strong pseudoprimality test and the Lucas pseudoprimality test to determine if n is a probable prime. (This has been implemented in Maple function *isprime.*) If it is, go to [2], else report that n is composite and go to [3].

[2] (Primality Proving – Elliptic Curve Proving) Use an elliptic curve test (e.g., the ECPP test) to verify whether or not n is indeed a prime. If it is, then report that n is prime, otherwise, report that n is composite.

[3] (Exit) Terminate the algorithm.

1.4 Intractable Number-Theoretic Problems

One of the most important features of number theory is that problems in number theory are easy to state but often very hard to solve. The following are just some of the problems that are still unsolved to date:

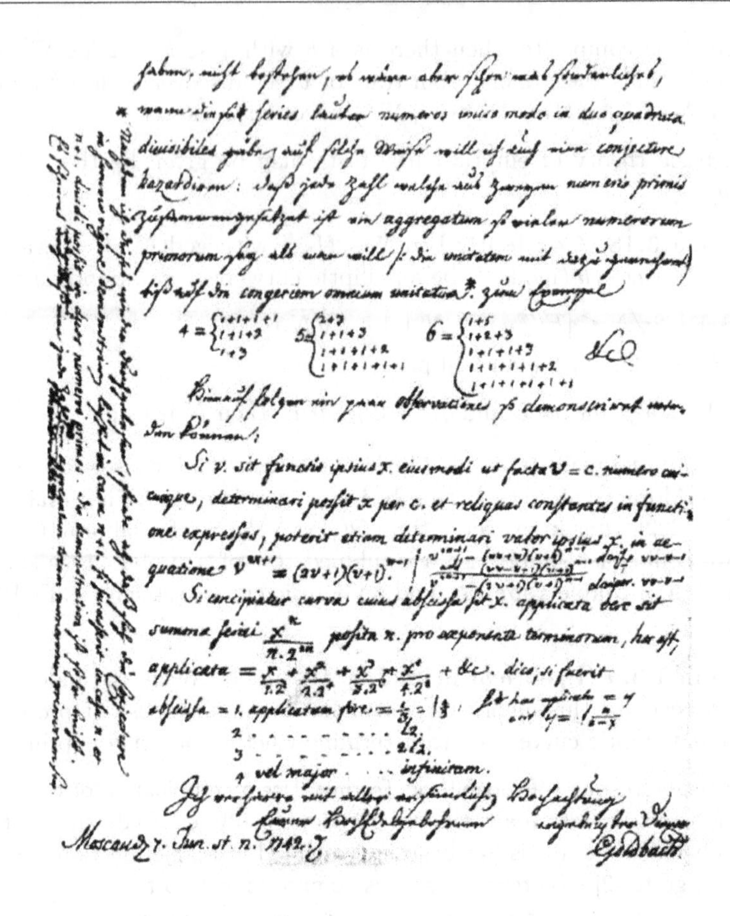

Figure 1.13. Goldbach's Letter to Euler

(1) The Goldbach Conjecture: In a letter to Euler (1707–1783), dated 7 June 1742 (see Figure 1.13), Goldbach (1690–1764) conjectured that *every even integer greater than 4 is the sum of two prime numbers*, such as $6 = 3+3$, $8 = 3+5$, $10 = 3+7 = 5+5$, $12 = 5+7$. $14 = 3+11$, $16 = 3+13 = 5+11$, $18 = 5+13 = 7+11$, etc. Euler believed it to be true but was unable to produce a proof during his lifetime. Despite much effort this conjecture remains unsolved to this day, and the best rigorous result remains that *every sufficiently large even integer can be written as the sum of a prime and a product of at most two primes* (e.g., $100 = 23 + 7 \cdot 11$), proved by the Chinese mathematician J. R. Chen (1933–1996) in 1966 (full proof provided in 1973 [65]), whereas the most recent numerical result is that *the conjecture is true up to* 10^{18}, verified by Silva on 25 April 2007 [289]. Faber and Faber Publishers offered a $1,000,000 prize to anyone who can prove/disprove Goldbach's conjecture between 20 March 2000 and

20 March 2002, but the prize went unclaimed and the conjecture remains open.

(2) Riemann Hypothesis: Let s be a complex variable (we write $s = \sigma + it$ with σ and t real, where $\sigma = \mathrm{Re}(s)$ is the real part of s, whereas $t = \mathrm{Im}(s)$ is the imaginary part of s). Then the Riemann ζ-function, $\zeta(s)$, is defined to be the sum of the following series

$$\zeta(s) = \sum_{n=1}^{\infty} n^{-s}. \tag{1.33}$$

Bernhard Riemann (1826–1866) in 1859 calculated the first five complex

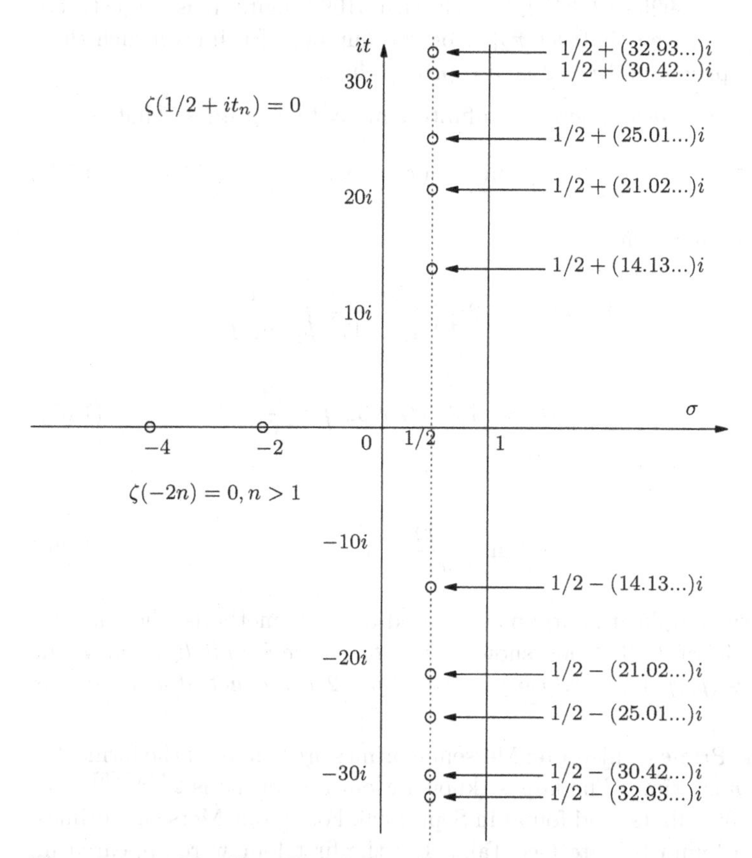

Figure 1.14. The First Five Zeros of ζ-Function above σ-Axis

zeros of the ζ-function above the σ-axis and found that they lie on the vertical line $\sigma = 1/2$. He then conjectured that *all the nontrivial (complex) zeros ρ of $\zeta(s)$ lying in the critical strip* $0 < \mathrm{Re}(s) < 1$ *must lie on the critical line* $\mathrm{Re}(s) = 1/2$. That is, $\rho = 1/2 + it$, where ρ denotes a nontrivial zero of $\zeta(s)$. Note that the real zeros are the trivial zeros, as $\zeta(-2n) = 0$ when $n = 1, 2, 3, \cdots$. The Riemann hypothesis may be true, it may also be false. At present, no-one knows whether or not it is true. The Clay Mathematics Institute of Cambridge in Massachusetts has named the Riemann hypothesis as one of its seven millennium Prize Problems [36]; the Board of Directors of the institute designated on 24 May 2000 $7 million prize fund for the solution to these problems, with $1 million allocated to each.

(3) Twin Prime Conjecture: Twin primes are defined to be pairs $(p,\ p+2)$, such that both p and $p+2$ are prime. For example, $(3,5)$, $(17,19)$ and $(1997,1999)$ are twin primes. The largest known twin primes, found in 2002, are $33218925 \cdot 2^{169690} \pm 1$, each with 51090 digits. It is conjectured but not yet proven that: let $\pi_2(x)$ be the number of primes p such that if $p \leq x$ is prime, and $p+2$ is also prime, then

(3-1) *A weak form:* There are infinitely many twin primes. That is,

$$\lim_{x \to \infty} \pi_2(x) = \infty. \tag{1.34}$$

(3-2) *A strong form:* Let

$$\begin{aligned} \mathrm{L}_2(x) &= 2 \prod_{p \geq 3} \frac{p(p-2)}{(p-1)^2} \int_2^x \frac{\mathrm{d}t}{\ln^2 t} \\ &\approx 1.320323632 \int_2^x \frac{\mathrm{d}t}{\ln^2 t} \end{aligned} \tag{1.35}$$

then

$$\lim_{x \to \infty} \frac{\pi_2(x)}{\mathrm{L}_2(x)} = 1. \tag{1.36}$$

Using very complicated arguments based on sieve methods, the Chinese mathematician J. R. Chen showed that *there are infinitely many pairs of integers $(p,\ p+2)$, with p prime and $p+2$ a product of at most two primes.*

(4) Mersenne Prime Conjecture: Mersenne primes are primes of the form $2^p - 1$, where p is prime. The largest known Mersenne prime is $2^{32582657} - 1$, with 9808358 digits, and found in Sept 2006. Forty-four Mersenne primes have been found to date (see Table 1.1; the first four were appeared in Euclid's *Elements* 2000 year ago). *It is conjectured but not yet proven that there are infinitely many Mersenne primes.* An interesting property of the Mersenne primes is that whenever $2^p - 1$ is a Mersenne prime, $2^{p-1}(2^p - 1)$ will be a perfect number (note that a number n is a perfect number if $\sigma(n) = 2n$; for example, 6,28,496,8128 are the first four perfect numbers). That is, n is a perfect number if and only if $n = 2^{p-1}(2^p - 1)$, where $2^p - 1$

No.	p	Digits	Time	No.	p	Digits	Time
1	2	1	–	2	3	1	–
3	5	2	–	4	7	3	–
5	13	4	1456	6	17	6	1588
7	19	6	1588	8	31	10	1772
9	61	19	1883	10	89	27	1911
11	107	33	1914	12	127	39	1876
13	521	157	1952	14	607	183	1952
15	1279	386	1952	16	2203	664	1952
17	2281	687	1952	18	3217	969	1957
19	4253	1281	1961	20	4423	1332	1961
21	9689	2917	1963	22	9941	2993	1963
23	11213	3376	1963	24	19937	6002	1971
25	21701	6533	1978	26	23209	6987	1979
27	44497	13395	1979	28	86243	25962	1982
29	110503	33265	1988	30	132049	39751	1983
31	216091	65050	1985	32	756839	227832	1992
33	859433	258716	1994	34	1257787	378632	1996
35	1398269	420921	1996	36	2976221	895932	1997
37	3021377	909526	1998	38	6972593	2098960	1999
39	13466917	4053946	2001	40	20996011	6320430	2003
41	24036583	7235733	2004	42	25964951	7816230	2005
43	30402457	9152052	2005	44	32582657	9808358	2006

Table 1.1. The 44 Known Mersenne Primes

is a Mersenne prime. This is the famous Euclid-Euler theorem which took about 2000 years to prove; the sufficient condition was established by Euclid in his *Elements*, but the necessary condition was established by Euler in work published posthumously. Again, we do not know if there are infinitely many perfect numbers, particularly, we do not know if there is an *odd perfect number*; what we know is that there is no odd perfect numbers up to 10^{300}, due to Brent, et al [52]. Whether or not there is an odd perfect number is one of the most important unsolved problems in all of mathematics.

(5) Arithmetic Progression of Consecutive Primes: An arithmetic progression of primes is defined to be the sequence of primes satisying:

$$p, p+d, p+2d, \cdots, p+(k-1)d \tag{1.37}$$

where p is the first term, d the common difference, and $p+(k-1)d$ the last term of the sequence. For example, the following are some sequences of the arithmetic progression of primes:

2				
2	3			
3	5	7		
5	11	17	23	
5	11	17	23	29

The longest arithmetic progression of primes is the following sequence with 23 terms: $56211383760397 + k \cdot 44546738095860$ with $k = 0, 1, \cdots, 22$. Thanks to Green and Tao [128] who proved that *there are arbitrary long arithmetic progressions of primes* (i.e., k can be any arbitrary large natural number), which enabled, among others, Tao to receive a Field Prize in 2006, an equivalent Nobel Prize for Mathematics. However, their result is not about consecutive primes; we still do not know if there are arbitrary long arithmetic progressions of consecutive primes, although Chowa [67] proved in 1944 that there exists an infinity of three consecutive primes of arithmetic progressions. Note that an arithmetic progression of consecutive primes is a sequence of consecutive primes in the progression. In 1967, Jones, Lal and Blundon [160] found an arithmetic progression of five consecutive primes $10^{10} + 24493 + 30k$ with $k = 0, 1, 2, 3, 4$. In the same year, Lander and Parkin [183] discovered six in an arithmetic progression $121174811 + 30k$ with $k = 0, 1, 2, 3, 4, 5$. The longest arithmetic progression of consecutive primes, discovered by Manfred Toplic in 1998, is $507618446770482 \cdot 193\# + x77 + 210k$, where $193\#$ is the product of all primes ≤ 193, i.e., $193\# = 2 \cdot 3 \cdot 5 \cdot 7 \cdots 193$, $x77$ is a 77-digit number 54538241683887582668189703590110659057865934764604873840781923513421103495579, and $k = 0, 1, 2, \cdots, 9$.

In this book, we are more interested in those number theoretic problems that are computationally intractable, since the security of modern public-key cryptography relies on the intractability of these problems. A problem is computationally intractable if it cannot be solved in polynomial-time. Thus, from a computational complexity point of view, any problem beyond $\mathcal{P}$ is intractable. There are, however, different types of intractable problems (see Figure 1.15).

(1) Provably intractable problems: Problems that are Turing computable but can be shown in $\mathcal{PS}$ ($\mathcal{P}$-$\mathcal{S}$pace), $\mathcal{NPS}$ ($\mathcal{NP}$-$\mathcal{S}$pace), $\mathcal{EXP}$ (exponential time) etc., of course outside $\mathcal{NP}$, are provably and certainly intractable. Note that although we do not know if $\mathcal{P} = \mathcal{NPS}$, we know $\mathcal{PS} = \mathcal{NPS}$.

(2) Presumably intractable problems: Problems in $\mathcal{NP}$ but outside of $\mathcal{P}$, particularly those problem in $\mathcal{NPC}$ ($\mathcal{NP}$-complete) such as the Travelling Salesman Problem, the Knapsack Problem, and the Satisfiability problem, are presumably intractable, since we do not know whether or not $\mathcal{P} = \mathcal{NP}$. If $\mathcal{P} = \mathcal{NP}$, then all problems in $\mathcal{NP}$ will no longer be intractable. However, it is more likely that $\mathcal{P} \neq \mathcal{NP}$. From a cryptographic

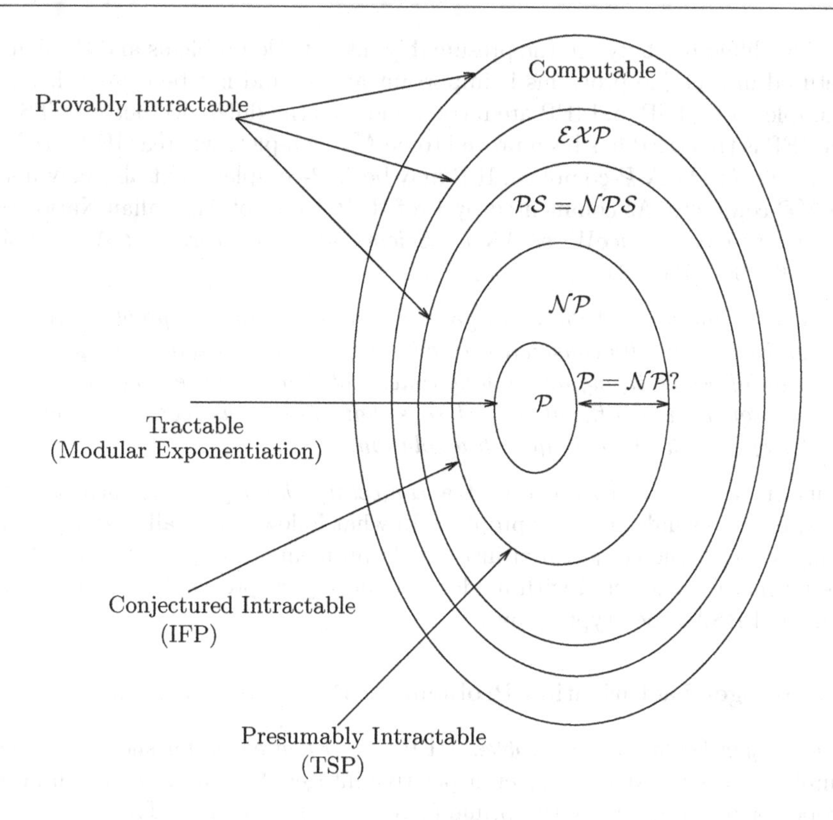

Figure 1.15. Tractable and Intractable Problems

point of view, it would be nice if encryption schemes can be designed to be based on some $\mathcal{NP}$-complete problems, since these types of schemes can be difficult to break. Experience, however, tells us that very few encryption schemes are based on $\mathcal{NP}$-complete problems.

(3) Conjectured intractable problems: By conjectured intractable problems we mean that the problems are currently in $\mathcal{NP}$-complete, but no-one can prove they must be in $\mathcal{NP}$-complete; they may be in $\mathcal{P}$ if efficient algorithms are invented for solving these problems. Typical problems in this category include the Integer Factorization Problem (IFP), the Discrete Logarithm Problem (DLP) and the Elliptic Curve Discrete Logarithm Problem (ECDLP). Again, from a cryptographic point of view, we are more interested in this type of intractable problems, and in fact, the IFP, DLP and ECDLP are essentially the only three intractable problems that are practical and widely used in commercial cryptography. For example, the most famous and widely used RSA cryptographic system relies its security on the intractability of the IFP problem.

The difference between the presumably intractable problems and the conjectured intractable problems is important and should not be confused. For example, both TSP and IFP are intractable, but the difference between TSP and IFP is that TSP has been proved to be $\mathcal{NP}$-complete whereas IFP is only *conjectured* to be $\mathcal{NP}$-complete. IFP may be $\mathcal{NP}$-complete, but also may not be $\mathcal{NP}$-complete. As commented by Prof R. P. Brent of Australian National University (see *ScienceWise@ANU – Science and Engineering at ANU*, Vol 2, No 6, 2005, Page 4):

> *One of the hardest things to prove is the difficulty of a problem. A problem is usually considered to be hard if no one can solve it despite a lot of people trying over a long time. But that's not the same thing as proving it is difficult. It just says that no-one has been clever or lucky enough to come up with a solution.*

Thus, it is not easy to prove that a *conjectured hard* problem such as the IFP problem is indeed a *hard* problem. In what follows, we shall examine and study some of the conjectured intractable problems in number theory, that are intimately connected with modern public-key cryptography, particularly RSA and RSA-type cryptography.

The Integer Factorization Problem (IFP)

The *Integer Factorization Problem* (IFP), or *IFP problem* for short, may be simply defined as follows. Given a positive integer N greater than 1, find a proper factor (not necessarily prime factor) $1 < f < N$ of N. That is,

$$\text{IFP}: \ \{N \in \mathbb{Z}^+_{>1}\} \xrightarrow[\text{hard}]{\text{find}} \{f \mid N \text{ with } 1 < f < N\}.$$

This problem is hard to solve. A further problem related to IFP is the Prime Factorization Problem (PFP), or *PFP Problem* for short, aiming at finding all the prime factors of N:

$$\text{PFP}: \ \{N \in \mathbb{Z}^+_{>1}\} \xrightarrow[\text{hard}]{\text{find}} \{N = \prod_{i=1}^{k} p_i^{\alpha_i} \text{ with } \alpha_i \in \mathbb{Z}^+, \ p_i \in \text{ Primes}\}.$$

An important theorem in Number Theory, called the *Fundamental Theorem of Arithmetic*, states that

Theorem 1.4.1 (Fundamental Theorem of Arithmetic). Any positive integer greater than 1 can be written uniquely in the following prime factorization form:

$$N = \prod_{i=1}^{k} p_i^{\alpha_i} \tag{1.38}$$

where $p_1 < p_2 < \cdots < p_k$ are primes and α_i are positive integers.

Thus, the PFP Problem is to find the prime factorization (1.38) of N. The theorem was essentially first proved by Euclid in his *Elements* [108], but the first full and correct proof is found in the *Disquisitiones Arithmeticae* by Gauss [118]. An interesting point to remark is that the standard proof of the existence and uniqueness of the theorem gives no hint of an *efficient* algorithm for computing the prime factorization of N. Obviously, if there is an efficient algorithm for IFP, then by recursively executing the algorithm, the PFP can be efficiently solved. Thus, IFP and PFP can be regarded as computationally equivalent problems:

$$\text{IFP} \overset{\mathcal{P}}{\Longleftrightarrow} \text{PFP}.$$

Note that primality testing and integer factorization are two different problems although they are intimately connected with each other, and often confused by many even very professional people including e.g., Bill Gates, who mentioned in this book *The Road Ahead* (page 265 in [117]) that

> The obvious mathematical breakthrough would be development of an easy way to *factor large prime* numbers.

In prime factorization, we do not factor a prime, but a composite, and in fact, we are trying to find prime factors *from* a large composite. If one wants to factor a prime, it can be easy, say, e.g.,

$$13 = (2 + 3i)(2 - 3i).$$

In fact, any prime number p satisfying $p \equiv 1 \pmod{4}$ can always be factored into Gaussian primes [18] in the form

$$p = -i(a + bi)(b + ai),$$

but that is not the problem we are interested in RSA cryptography.

Note also that there are many integer factorization algorithms, from ancient to modern. The only problem is that we do not have an efficient factoring algorithm. The fastest factoring algorithm, the Number Field Sieve (NFS), runs in expected time $\mathcal{O}(\exp(c(\log N)^{1/3}(\log\log n)^{2/3})))$, where $c = (64/9)^{1/3}$ if a general version of NFS is used for an arbitrary integer N, and $c = (32/9)^{1/3}$ if a special version of NFS is used for a special form of numbers such as the Fermat numbers $F_n = 2^{2^n} + 1$, $n \geq 5$. Again, many people are confused with *efficient algorithms* and *algorithms* (by definition, an algorithm is an effective procedure, and an algorithm is efficient if it runs in polynomial-time). For example, the later Professor Richard Feynman, the 1965 Noble Prize Laureate in Physics, mentioned the following statements in his book *Feynman Lectures on Computation* (page 91 in [112]):

> Another similar problem deals with factorization: I gave you a number m, and tell you that it is the product of two primes, $m = pq$. You have to find p and q. This problem *does not have an effective procedure* as yet, and it in fact forms the basis of a coding system.

Feynman was wrong! As we have just mentioned, there are many effective procedures (i.e., algorithms), from ancient to modern, for integer factorization, only none of them is efficient. By definition, an effective procedure is an algorithm. Thus, *Effective procedure* and *efficient algorithm* are two different concepts; the former is connected to computability, whereas the latter is related to complexity. To develop an efficient (i.e., polynomial-time) factoring algorithm is one of the most important research topics in computational number theory.

The Root Finding Problem (RFP)

The k-th Root Finding Problem (RFP), or *RFP Problem* for short, may be defined as follows:

$$k\text{RFP}: \ \{k, N, y \equiv x^k \pmod{N}\} \xrightarrow[\text{hard}]{\text{find}} \{x \equiv \sqrt[k]{y} \pmod{N}\}.$$

If the prime factorization of N is known, one can compute the Euler function $\phi(N)$ and solve the linear Diophantine equation $ku - \phi(N)v = 1$ in u and v, and the computation $x \equiv y^u \pmod{N}$ gives the required value. Thus, if IFP can be solved in polynomial-time, then RFP can also be solved in polynomial-time:

$$\text{IFP} \stackrel{\mathcal{P}}{\Longrightarrow} \text{RFP}.$$

The security of RSA relies on the intractability of IFP, and also on RFP; if any one of the problems can be solved in polynomial-time, RSA can be broken in polynomial-time.

The Quadratic Residuosity Problem (QRP)

Let $N \in \mathbb{Z}^+_{>1}$, $\gcd(y, N) = 1$. Then y is a quadratic residue modulo N, denoted by $y \in \text{QR}_\text{N}$, if the quadratic congruence

$$x^2 \equiv y \pmod{N},$$

has a solution in x. If the congruence has no solution in x, then y is a quadratic nonresidue modulo N, denoted by $y \in \overline{\text{QR}}_N$. The Quadratic Residuosity Problem (QRP), or the *QRP Problem* for short, is to decide whether or not $y \in \text{QR}_N$:

$$\text{QRP}: \quad \{n \in \mathbb{Z}^+_{>1}, x^2 \equiv y \pmod{N}\} \xrightarrow[\text{hard}]{\text{decide}} \{y \in \text{QR}_N\}.$$

If N is prime, or the prime factorization of N is known, then QRP can be solved simply by evaluating the Legendre symbol $L(y, N)$. If n is not a prime then one evaluates the Jacobi symbol $J(y, N)$ which, unfortunately, does not reveal if $y \in \text{QR}_N$, i.e., $J(y, N) = 1$ does not imply $y \in \text{QR}_N$ (it does if N

is prime). For example, $L(15, 17) = 1$, so $x^2 \equiv 15 \pmod{17}$ is soluble, with $x = \pm 21$ being the two solutions. However, although $J(17, 21) = 1$ there is no solution for $x^2 \equiv 17 \pmod{21}$. Thus, when N is composite, the only way to decide whether or not $y \in \mathrm{QR}_N$ is to factor N. Thus, if IFP can be solved in polynomial-time, QRP can also be solved in polynomial-time:

$$\text{IFP} \overset{\mathcal{P}}{\Longrightarrow} \text{QRP}.$$

The security of Goldwasser-Micali probabilistic encryption scheme [124] is based on the intractability of QRP.

The Square Root Problem (SQRT)

Let $y \in \mathrm{QR}_N$. The SQRT is to find an x such that

$$x^2 \equiv y \pmod{N} \quad \text{or} \quad x \equiv \sqrt{y} \pmod{N}.$$

That is,

$$\text{SQRT}: \quad \{N \in \mathbb{Z}^+_{>1}, y \in \mathrm{QR}_n, y \equiv x^2 \pmod{N}\} \xrightarrow[\text{hard}]{\text{find}} \{x\}.$$

When N is prime, the *SQRT problem* can be solved in polynomial-time. However, when N is composite one needs to factor N first. Thus, if IFP can be solved in polynomial-time, SQRT can also be solved in polynomial-time:

$$\text{IFP} \overset{\mathcal{P}}{\Longrightarrow} \text{QRP}.$$

On the other hand, if SQRT can be solved in polynomial-time, IFP can also be solved in polynomial-time:

$$\text{SQRT} \overset{\mathcal{P}}{\Longrightarrow} \text{IFP}.$$

That is,

$$\text{SQRT} \overset{\mathcal{P}}{\Longleftrightarrow} \text{IFP}.$$

It is precisely this intractability of SQRT that Rabin used to construct his cryptosystem in 1979 [252].

The Discrete Logarithm Problem (DLP)

Let $x, y, N, k \in \mathbb{Z}^+$. The *DLP problem* for multiplicative group $\mathbb{Z}^*_N$ may be described as follows:

$$\text{DLP}: \quad \{n \in \mathbb{Z}^+_{>1}, x \in \mathbb{Z}^+, y \equiv x^k \pmod{N}\} \xrightarrow[\text{hard}]{\text{find}} \{k\}.$$

DLP is believed to have about the same degree of difficulty as IFP; since methods for IFP are always, subject to some modifications, applicable to DLP.

$$\text{IFP} \overset{\mathcal{P}}{\Longrightarrow} \text{DLP}.$$

Several well known cryptosystems, such as the Diffie-Hellman-Merkle key-exchange scheme, the ElGammal systems and the DSA/DSS, base their security on the intractability of DLP.

The Elliptic Curve Discrete Logarithm Problem (ECDLP)

The *ECDLP problem* for the elliptic curve group $E(\mathbb{F}_p)$ is an extension of the DLP problem defined previously. Let $E/\mathbb{F}_p$ be an elliptic curve over a finite field $\mathbb{F}_p$:

$$E: \ y^2 \equiv x^3 + ax + b \pmod{p},$$

$E(\mathbb{F}_p)$ the set of points on $E/\mathbb{F}_p$, and $P, Q \in E(\mathbb{F}_p)$. Then the ECDLP is to find the integer k (assuming that k exists) such that $Q \equiv kP \pmod{p}$; that is,

$$\text{ECDLP}: \quad \{P \in E(\mathbb{F}_p), p \in \text{Primes}, Q \equiv kP \pmod{p}\} \xrightarrow[\text{hard}]{\text{find}} \{k\}.$$

ECDLP is believed to be harder than DLP and is the security basis for various elliptic curve cryptosystems [337]. However, from a quantum computation point of view, all IFP, DLP and ECDLP are quantum polynomial-time equivalent; they all can be solved in polynomial-time equivalent on a quantum computer.

Modular Polynomial Root Finding Problem (MPRFP)

There are also many other hard number-theoretic problems. For example, it is easy to compute the integer roots of a polynomial in one variable over $\mathbb{Z}$:

$$p(x) = 0 \tag{1.39}$$

but the following *modular polynomial root finding problem (MPRFP)*, or the *MPRFP problem* for short, can be hard:

$$p(x) \equiv 0 \pmod{N}, \tag{1.40}$$

which aims at finding integer roots (solutions) of the modular polynomial in one variable. This problem can, of course, be extended to find integer roots (solutions) of the modular polynomial in several variables as follows:

$$p(x, y, \cdots) \equiv 0 \pmod{N}. \tag{1.41}$$

Coppersmith [78] developed a powerful method to find all small solutions x_0 of the modular polynomial equations in one or two variables of degree δ using the lattice reduction algorithm LLL [188] (we shall discuss Coppersmith method later). Of course, for LLL to be run in reasonably amount of time in finding such x_0's, the values of δ cannot be big.

Shortest Vector Problem (SVP)

Problems related to lattices are also often hard to solve. Let $\mathbb{R}^n$ denote the space of n-dimensional real vectors $a = \{a_1, a_2, \cdots, a_n\}$ with usual dot product $a \cdot b$ and Euclidean Norm or length $||a|| = (a \cdot a)^{1/2}$. $\mathbb{Z}^n$ is the set of vectors in $\mathbb{R}^n$ with integer coordinates. If $A = \{a_1, a_2, \cdots, a_n\}$ is a set of linear independent vectors in $\mathbb{R}^n$, then the set of vectors

$$\left\{ \sum_{i=1}^{n} k_i a_i \; : \; k_1, k_2, \cdots, k_n \in \mathbb{Z} \right\} \tag{1.42}$$

is a lattice in $\mathbb{R}^n$, denoted by $L(A)$ or $L(a_1, a_2, \cdots, a_n)$. A is called a basis of the lattice. A set of vectors in $\mathbb{R}^n$ is a n-dimensional lattice if there is a basis V of n linear independent vectors such that $L = L(V)$. If $A = \{a_1, a_2, \cdots, a_n\}$ is a set of vectors in a lattice L, then the length of the set A is defined by $\max(||a_i||)$. A fundamental theorem, due to Minkowski, is as follows.

Theorem 1.4.2 (Minkowski). There is a universal constant γ, such that for any lattice L of dimension n, $\exists v \in L$, $v \neq 0$, such that

$$||v|| = \gamma \sqrt{n} \det{(L)}^{1/n}. \tag{1.43}$$

The determinant $\det(L)$ of a lattice is the volume of the n-dimensional fundamental parallelepiped, and the absolute constant γ is known as Hermite's constant.

A natural problem concerned with lattices is the *shortest vector problem (SVP)*, or the *SVP problem* for short:

Find the shortest non-zero vector in a high dimensional lattice.

Minkowski's theorem is just an existence-type theorem and offers no clue on how to find a short or the shortest vector non-zero vector in a high dimensional lattice. There is no efficient algorithm for finding the shortest non-zero vector, or finding an approximate short non-zero vector. The lattice reduction algorithm LLL [188] can be used to find short vectors, but it is not effective in finding short vectors when the dimension n is large, say, for example, $n \geq 100$. This allows lattices to be used in the design of cryptographic systems, and in fact, several cryptographic systems, such as NTRU [144] and the Ajtai-Dwork system [9], are based on the intractability of finding the shortest non-zero vector in a high dimensional lattice.

So the computationally intractable mathematical problems are two-edged swords; they have both fortunate and unfortunate edges. On the one hand, it is unfortunate for mathematicians, computer scientists and cryptanalysts that these problems cannot be solved in polynomial-time at least at the present time. On the other hand, it is very fortunate for cryptographers that these intractable problems can be used to construct secure cryptographic systems that is unbreakable in polynomial-time. For example, the most famous RSA cryptographic system is based on the intractability of the integer factorization problem (IFP). However, we must keep this in mind that these computationally intractable mathematical problems may be solvable in polynomial-time, or more seriously, some of the problems, for example, the IFP problem, might have already been solved in polynomial-time secretly.

1.5 Chapter Notes and Further Reading

The theory of computability, complexity and intractability is very important in cryptography and cryptanalysis. The following books provide a good survey on the subject: Garey and Johnson [116], Hopcroft, Motwani and Ullman [145], Lewis and Papadimitrou [191], and Sipser [304]. In particular, the seminal papers by Church [69], Turing [316], Cook [72] and karp [164] are the classical in the field.

RSA is a number-theoretic cryptographic system, so to understand RSA, particularly to work on the cryptanalysis of RSA, it is necessary to have a good background in number theory, particularly in computational number theory. There are many excellent books in number theory with various levels of difficulty. For a general coverage and a good introduction to elementary number theory, see Baker [18], Davenport [93], Everest and Ward [110], Hardy and Wright [138], Ireland and Rosen [153], LeVeque [190], Niven et al [231], Nathanson [227], or Silverman [294]. For analytic number theory, see Apostol [13], Stopple [312], and Bateman and Diamond [20]. The books by Ingham [152], Edwards [105], Patterson [235] present a good treatment on the prime distributions and the theory of the Riemann zeta function. For algebraic number theory, Mollin [215], and Alaca and Williams [10] are strongly recommended. For computational and/or algorithmic number theory, readers should consult e.g., the following books for more information: Bach and Shallit [17], Bressoud and Wagon [54], Cohen [71], Crandall and Pomerance [90], Riesel [260], Shoup [309], and Yan [335]. Finally, background information on lattice theory or more generally the geometry of numbers can be found in the classical writing by Cassels [64].

2. RSA Public-Key Cryptography

The increased use of shared communications channels, particularly wireless and local area networks (LAN's), leads to greater connectivity, but also to a much greater opportunity to intercept data and forge messages, $\cdots$ The only practical way to maintain privacy and integrity of information is by using public-key cryptography.

PETER WEGNER
Professor of Computer Science, Brown University

2.1 Introduction

Cryptography (from the Greek *Kryptós*, "hidden" or "secret", and gráphein, "writing") is the study of the processes of encryption (mapping the original message, called the plaintext, into a secret form, called the the ciphertext, using the encryption key), and decryption (inverting the ciphertext back to the plaintext, using the corresponding decryption key), in such a way that only the intended recipients can decrypt and read the original messages.

$$\text{Cryptograpgy} \stackrel{\text{def}}{=} \text{Encryption} \oplus \text{Decryption}$$

The methods of encryption are often also called ciphers. Cryptanalysis (from the Greek *Kryptós* and *analýein*, "loosing"), on the other hand, is the study of breaking the encryptions without the knowledge of the key:

$$\text{Cryptanalysis} \stackrel{\text{def}}{=} \text{Cryptanalytic Attacks on Encryptions}$$

Cryptology (from the Greek *Kryptós* and lógos, "word") consists of both cryptography and cryptanalysis:

$$\text{Cryptology} \stackrel{\text{def}}{=} \text{Cryptography} \oplus \text{Cryptanalysis}$$

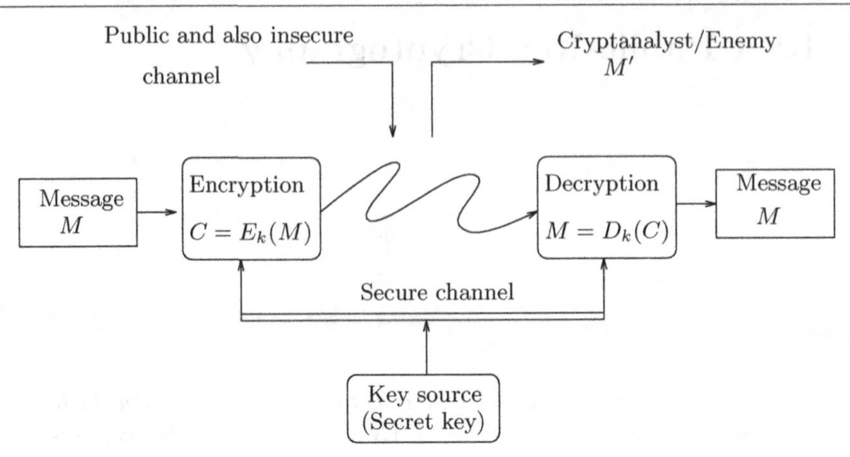

Figure 2.1. Cryptography and Cryptanalysis

The idea of encryption, decryption, cryptanalysis and secure communications over an insecure channel, usually a computer network particularly the Internet, can be depicted as in Figure 2.1. Throughout the book, we shall assume that Bob sends a message to Alice, but Eve wants to cryptanalyze the message:

$$\text{Bob} \xrightarrow[\underset{\text{Eve}}{\downarrow}]{\text{Message}} \text{Bob}. \tag{2.1}$$

Modern cryptography, however, is the study of the mathematical systems of encryption and decryption, to solve the security, particularly the network security problems as follows:

(1) *Confidentiality* or *privacy*: To stop Eve to understand Bob's message to Alice even if she can intercept and get the message.

(2) *Integrity*: To make sure that Bob's message has not been modified by Eve.

(3) *Authentication* or *authorization*: To make sure the message received by Alice is indeed from Bob, not from Eve.

(4) *Non-repudiation*: To stop Bob later to deny the sending of his message. Non-repudiation is particularly important in electronic commerce since we need to make sure that a consumer cannot deny the authorization of a purchase. It must be noted that however, in some applications such as in electronic voting, the non-repudiation feature should, in fact, be avoided, since the voter does not want to disclose the authorization of a vote regardless whether of not he actually did the vote.

Such a mathematical system is called the *cryptographic system*, or *cryptosystems* for short.

Definition 2.1.1. A conventional *secret-key cryptosystem* (or *secret-key encryption*, or *secret-key cipher*) $\mathcal{S}$ may be formally defined as follows (depicted in Figure 2.2):

$$\mathcal{S} = (\mathcal{M}, \mathcal{C}, \mathcal{K}, M, C, k, E, D) \tag{2.2}$$

where

(1) $\mathcal{M}$ is the set of plaintexts, called the plaintext space.
(2) $\mathcal{C}$ is the set of cipherexts, called the ciphertext space.
(3) $\mathcal{K}$ is the set of keys, called the key space.
(4) $M \in \mathcal{M}$ is a piece of plaintext.
(5) $C \in \mathcal{C}$ is a piece of ciphertext.
(6) $k \in \mathcal{K}$ is the key for both encryption and decryption.
(7) E is the encryption function

$$E_k : M \mapsto C$$

where $M \in \mathcal{M}$ maps to $C \in \mathcal{C}$, using the key k, such that

$$C = E_k(M) \tag{2.3}$$

(8) D is the decryption function

$$D_k : C \mapsto M$$

where $C \in \mathcal{C}$ maps to $M \in \mathcal{M}$, using the *same* key k again such that

$$M = D_k(C) \tag{2.4}$$

satisfying

$$E_k D_k = 1 \text{ and } D_k(C) = D_k(E_k(M)) = M. \tag{2.5}$$

Cryptanalysis, on the other hand, is the study of the *cryptanalytic attacks* on cryptosystems, aiming at breaking the cryptosystems without using/knowing the keys, but according to the *Kerckhoff principle*, the cryptanalyst who wants to break the cryptosystem knows the cryptosystem. For example, the following is a ciphertext presented by Édouard Lucas at the 1891 meeting of the French Association for Advancement of Science (see page 388 of Williams [332]), based on Étienne Bazeries' cylindrical cryptography (see pages 244–250 of Kahn [161]); it has never been decrypted, and hence is suitable as a good challenge to the interested reader:

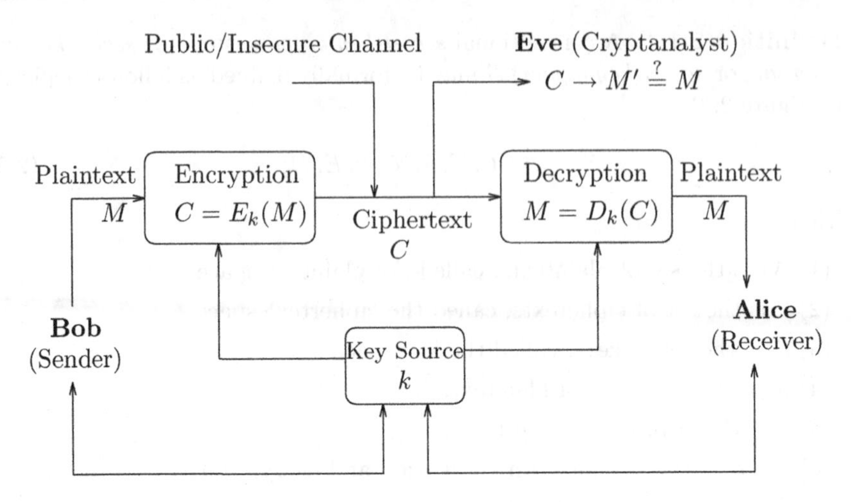

Figure 2.2. Secret-Key Cryptography

XSJOD	PEFOC	XCXFM	RDZME
JZCOA	YUMTZ	LTDNJ	HBUSQ
XTFLK	XCBDY	GYJKK	QBSAH
QHXPE	DBMLI	ZOYVQ	PRETL
TPMUK	XGHIV	ARLAH	SPGGP
VBQYH	TVJYJ	NXFFX	BVLCZ
LEFXF	VDMUB	QBIJV	ZGGAI
TRYQB	AIDEZ	EZEDX	KS

The *security* or the *unbreakability* of any cryptographic system is of paramount importance. There are several different types of security measures for a cryptographic system:

(1) *Unconditionally secure*: A cryptosystem is unconditionally secure if a cryptanalyst cannot determine how to find the plaintext M regardless of how much ciphertext C and computer time/resources he has available to him. A one-time pad (OTP) can be shown to be unconditionally secure, as the key is used only for one time (i.e., there are at least as many keys as the plaintexts), the key string is a random string, and the key size is at least as long as the plaintext string. Unconditional security for cryptosystems is called *perfect secrecy*, or *information-theoretic security*. A cryptosystem $\mathcal{S}$ is *unconditionally unbreakable* if $\mathcal{S}$ is unconditionally secure. In general, cryptosystems do not offer perfect secrecy, in particular, public-key cryptosystems, such as the RSA cryptosystem described in next sections, cannot be unconditionally secure/breakable since once

a ciphertext C is given, its corresponding plaintext M can in principle be recovered by computing all possible plaintexts until C is obtained, an attack called forward search, which will be discussed later. Nevertheless, unconditionally unbreakable cryptosystem exists; it was first proved by Shannon in his 1949 seminar paper in modern cryptography "Communication Theory of Secrecy Systems" [285]. Thus the prominent English mathematician J. E. Littlewood (1885–1977) commented:

The legend that every cipher is breakable is of course absurd, though still widespread among people who should know better.

(2) *Computationally secure*: A cryptosystem $\mathcal{S}$ is computationally secure or *polynomially secure* if a cryptanalyst cannot decrypt C to get M in polynomial-time (or space). A cryptosystem $\mathcal{S}$ is *computationally unbreakable*, if it is unbreakable in polynomial-time, that is, it is computationally secure. According to the Cook-Karp thesis, any problem that can not be solved in polynomial-time is *computationally infeasible*, thus, if the cryptanalytic attack on a cryptosystem is computationally infeasible, then the cryptosystem is computationally secure and computationally unbreakable. There are several types of computationally security:

(2-1) *Provably secure*: A cryptosystem $\mathcal{S}$ is *provably secure* if the difficulty of breaking it can be shown to be essentially as difficult as solving a well-known and supposedly difficult mathematical problems such as the integer factorization problem IFP or the discrete logarithm problem DLP. For example, the Rabin cryptosystem described later is provably secure, as the security of the Rabin cryptosystem is equivalent to the IFP problem.

(2-2) *Practical/conjectured secure*: A cryptosystem $\mathcal{S}$ is *practical secure* if the breaking of the system $\mathcal{S}$ is *conjectured* as difficult as solving a well-known and supposedly difficult mathematical problems such as the integer factorization problem IFP or the discrete logarithm problem DLP. For example, breaking the most popular public-key cryptosystem RSA is conjectured as hard as solving the IFP problem, but so far this has never been proved or disproved. Most of the public-key and secret-key cryptosystems in current use are in this type.

There are several types of possible cryptanalytic attacks on a cryptosystem $\mathcal{S}$, depending on what information the cryptanalyst might already have regarding $\mathcal{S}$:

(1) *Ciphertext-only attack*: Only a piece of ciphertext C is known to the cryptanalyst whose goal is to find the corresponding plaintext M and/or the key k. This is the most difficult type of attack; any cryptosystem vulnerable to this type of attack is considered to be *completely* insecure.

(2) *Known-plaintext attack*: The cryptanalyst has a piece of plaintext M and the corresponding ciphertext C. The goal is the find the key k so that other ciphertexts using the same encryption/key may be decrypted.

(3) *Chosen-plaintext attack*: The cryptanalyst has gained temporary access to the encryption machinery, so he can choose a piece of plaintext M and construct the corresponding ciphertext C. The goal here is to find the key k.

(4) *Chosen-ciphertext attack*: The cryptanalyst has gained temporary access to the decryption machinery, so can choose a piece of ciphertext C and construct the corresponding plaintext M. The goal here is also to find the key k.

A good cryptosystem should resist all of these types of attacks, so that it is impossible for a cryptanalysis to get the key k or to find the plaintext M in polynomial-time.

Remark 2.1.1. Public-key cryptosystems, such as the RSA cryptosystem described in the next sections, give rise to the chosen-ciphertext attack, since the cryptanalyst may specify/obtain some ciphertext using the public-key and learn the corresponding plaintext. In fact, all public-key cryptographic systems are vulnerable to a chosen-ciphertext attack, which, however, can be avoided by adding appropriate redundancy or randomness (padding or salting) prior to encryption.

2.2 Public-Key Cryptography

Surprisingly, *public-key cryptography*, or *asymmetric key cryptography* (see Figure 2.3), is almost the same (although the idea is different) as the *secret-key cryptography*, or *symmetric key cryptography*, except that the keys k for encryption and decryption are different. That is, we need two keys, e_k and d_k, such that e_k is used for encryption and d_k for decryption, respectively. As e_k is only used for encryption, it can be made public; only d_k must be kept a secret for decryption. To distinguish public-key cryptosystems from secret-key cryptosystems, e_k is called the *public key*, and d_k the *private key*; only the key used in secret-key cryptosystems is called the *secret key*. Remarkably enough, secret-key cryptography has a very long history, almost as long as our human civilization; whereas public-key cryptography has a rather short history. In fact the official date of birth of public-key cryptography is 1976, when Diffie and Hellman, then both at Stanford University, published their seminal paper *New Directions in Cryptography* [101] (see the first page of the paper in Figure 2.4). It is in this seminal paper that they *first* publicly proposed the completely new idea of public-key cryptography as well as digital signatures. Although Diffie and Hellman did not have a practical implementation of their idea, they did propose [101] an alternative key-exchange scheme over the

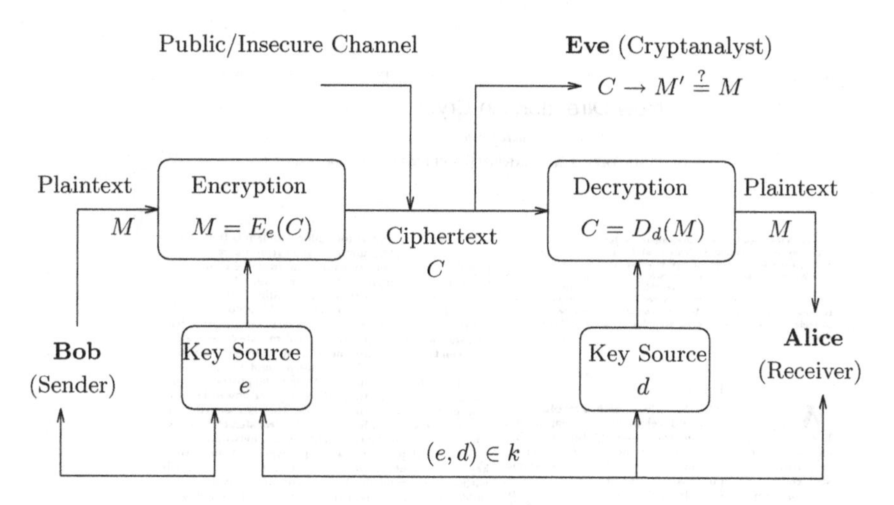

Figure 2.3. Public-Key Cryptography

insecure channel, based on the intractability of the DLP problem and using some of the ideas proposed earlier (although published later) by Merkle [208] (published in 1978, but submitted in 1975; see the first page of this paper in Figure 2.5).

Shortly after the publication of Diffie and Hellman's paper, Rivest, Shamir and Adleman, then all at Massachusetts Institute of Technology (MIT), proposed a first workable and practical public-key cryptosystem in in 1977 [262] (see the first page of the paper in Figure 2.6). The system is now known as RSA; it was first made public to the world and became famous probably because Gardner's 1978 paper in Scientific American [115].

It is interesting to note that the British cryptographers Ellis, Cocks and Williamson at the UK Government's Communications-Electronics Security Group/Government Communications Headquarters (CESG/GCHQ) also claimed that they secretly discovered the public-key encryption years before the US scientists. There are of course two different universes of cryptography: public (particularly for people working in academic institutions) and secret (particularly for people working for militaries and governments). Ellis-Cocks-Williamson certainly deserved some credit for their contribution to the development of public-key cryptography. It should be noted that Hellman and his colleagues not only invented the public-key encryption, but also the digital signatures which had not been mentioned in any of Ellis-Cocks-Williamson's documents/papers.

644 IEEE TRANSACTIONS ON INFORMATION THEORY, VOL. IT-22, NO. 6, NOVEMBER 1976

New Directions in Cryptography

Invited Paper

WHITFIELD DIFFIE AND MARTIN E. HELLMAN, MEMBER, IEEE

***Abstract*—Two kinds of contemporary developments in cryptography are examined. Widening applications of teleprocessing have given rise to a need for new types of cryptographic systems, which minimize the need for secure key distribution channels and supply the equivalent of a written signature. This paper suggests ways to solve these currently open problems. It also discusses how the theories of communication and computation are beginning to provide the tools to solve cryptographic problems of long standing.**

I. Introduction

WE STAND TODAY on the brink of a revolution in cryptography. The development of cheap digital hardware has freed it from the design limitations of mechanical computing and brought the cost of high grade cryptographic devices down to where they can be used in such commercial applications as remote cash dispensers and computer terminals. In turn, such applications create a need for new types of cryptographic systems which minimize the necessity of secure key distribution channels and supply the equivalent of a written signature. At the same time, theoretical developments in information theory and computer science show promise of providing provably secure cryptosystems, changing this ancient art into a science.

The development of computer controlled communication networks promises effortless and inexpensive contact between people or computers on opposite sides of the world, replacing most mail and many excursions with telecommunications. For many applications these contacts must be made secure against both eavesdropping and the injection of illegitimate messages. At present, however, the solution of security problems lags well behind other areas of communications technology. Contemporary cryptography is unable to meet the requirements, in that its use would impose such severe inconveniences on the system users, as to eliminate many of the benefits of teleprocessing.

Manuscript received June 3, 1976. This work was partially supported by the National Science Foundation under NSF Grant ENG 10173. Portions of this work were presented at the IEEE Information Theory Workshop, Lenox, MA, June 23–25, 1975 and the IEEE International Symposium on Information Theory in Ronneby, Sweden, June 21–24, 1976.

W. Diffie is with the Department of Electrical Engineering, Stanford University, Stanford, CA, and the Stanford Artificial Intelligence Laboratory, Stanford, CA 94305.

M. E. Hellman is with the Department of Electrical Engineering, Stanford University, Stanford, CA 94305.

The best known cryptographic problem is that of privacy: preventing the unauthorized extraction of information from communications over an insecure channel. In order to use cryptography to insure privacy, however, it is currently necessary for the communicating parties to share a key which is known to no one else. This is done by sending the key in advance over some secure channel such as private courier or registered mail. A private conversation between two people with no prior acquaintance is a common occurrence in business, however, and it is unrealistic to expect initial business contacts to be postponed long enough for keys to be transmitted by some physical means. The cost and delay imposed by this key distribution problem is a major barrier to the transfer of business communications to large teleprocessing networks.

Section III proposes two approaches to transmitting keying information over public (i.e., insecure) channels without compromising the security of the system. In a *public key cryptosystem* enciphering and deciphering are governed by distinct keys, E and D, such that computing D from E is computationally infeasible (e.g., requiring 10^{100} instructions). The enciphering key E can thus be publicly disclosed without compromising the deciphering key D. Each user of the network can, therefore, place his enciphering key in a public directory. This enables any user of the system to send a message to any other user enciphered in such a way that only the intended receiver is able to decipher it. As such, a public key cryptosystem is a multiple access cipher. A private conversation can therefore be held between any two individuals regardless of whether they have ever communicated before. Each one sends messages to the other enciphered in the receiver's public enciphering key and deciphers the messages he receives using his own secret deciphering key.

We propose some techniques for developing public key cryptosystems, but the problem is still largely open.

Public key distribution systems offer a different approach to eliminating the need for a secure key distribution channel. In such a system, two users who wish to exchange a key communicate back and forth until they arrive at a key in common. A third party eavesdropping on this exchange must find it computationally infeasible to compute the key from the information overheard. A possible solution to the public key distribution problem is given in Section III, and Merkle [1] has a partial solution of a different form.

A second problem, amenable to cryptographic solution, which stands in the way of replacing contemporary busi-

Figure 2.4. The First Page of Diffie and Hellman's Paper

The implementation of public-key cryptosystems is based on *trapdoor one-way functions.*

Definition 2.2.1. Let S and T be finite sets. A *one-way function*

$$f: S \to T \tag{2.6}$$

is an invertible function satisfying

(1) f is easy to compute, that is, given $x \in S$, $y = f(x)$ is easy to compute.

(2) f^{-1}, the inverse function of f, is difficult to compute, that is, given $y \in T$, $x = f^{-1}(y)$ is difficult to compute.

Programming Techniques — S. L. Graham, R. L. Rivest, Editors

Secure Communications Over Insecure Channels

Ralph C. Merkle
Department of Electrical Engineering and Computer Sciences
University of California, Berkeley

According to traditional conceptions of cryptographic security, it is necessary to transmit a key, by secret means, before encrypted messages can be sent securely. This paper shows that it is possible to select a key over open communications channels in such a fashion that communications security can be maintained. A method is described which forces any enemy to expend an amount of work which increases as the square of the work required of the two communicants to select the key. The method provides a logically new kind of protection against the passive eavesdropper. It suggests that further research on this topic will be highly rewarding, both in a theoretical and a practical sense.

Key Words and Phrases: security, cryptography, cryptology, communications security, wiretap, computer network security, passive eavesdropping, key distribution, public key cryptosystem

CR Categories: 3.56, 3.81

General permission to make fair use in teaching or research of all or part of this material is granted to individual readers and to nonprofit libraries acting for them provided that ACM's copyright notice is given and that reference is made to the publication, to its date of issue, and to the fact that reprinting privileges were granted by permission of the Association for Computing Machinery. To otherwise reprint a figure, table, other substantial excerpt, or the entire work requires specific permission as does republication, or systematic or multiple reproduction.

Author's address: Dept. of Electrical Engineering, Stanford University, Stanford CA, 94305.

©1978 ACM 0001-0782/78/0400-0294 $00.75

Communications of the ACM — April 1978, Volume 21, Number 4

Figure 2.5. The First Page of Merkle's Paper

(3) f^{-1} is easy to compute when a trapdoor (i.e., a secret string of information associated with the function) becomes available.

A function f satisfying only the first two conditions is also called a one-to-one one-way function. If f satisfies further the third condition, it is called a *trapdoor one-way function.*

Programming Techniques — S.L. Graham, R.L. Rivest* Editors

A Method for Obtaining Digital Signatures and Public-Key Cryptosystems

R. L. Rivest, A. Shamir, and L. Adleman
MIT Laboratory for Computer Science
and Department of Mathematics

An encryption method is presented with the novel property that publicly revealing an encryption key does not thereby reveal the corresponding decryption key. This has two important consequences:
(1) Couriers or other secure means are not needed to transmit keys, since a message can be enciphered using an encryption key publicly revealed by the intended recipient. Only he can decipher the message, since only he knows the corresponding decryption key.
(2) A message can be "signed" using a privately held decryption key. Anyone can verify this signature using the corresponding publicly revealed encryption key. Signatures cannot be forged, and a signer cannot later deny the validity of his signature. This has obvious applications in "electronic mail" and "electronic funds transfer" systems. A message is encrypted by representing it as a number M, raising M to a publicly specified power e, and then taking the remainder when the result is divided by the publicly specified product, n, of two large secret prime numbers p and q. Decryption is similar; only a different, secret, power d is used, where e * d ≡ 1(mod (p − 1) * (q − 1)). The security of the system rests in part on the difficulty of factoring the published divisor, n.

Key Words and Phrases: digital signatures, public-key cryptosystems, privacy, authentication, security, factorization, prime number, electronic mail, message-passing, electronic funds transfer, cryptography.

CR Categories: 2.12, 3.15, 3.50, 3.81, 5.25

General permission to make fair use in teaching or research of all or part of this material is granted to individual readers and to nonprofit libraries acting for them provided that ACM's copyright notice is given and that reference is made to the publication, to its date of issue, and to the fact that reprinting privileges were granted by permission of the Association for Computing Machinery. To otherwise reprint a figure, table, other substantial excerpt, or the entire work requires specific permission as does republication, or systematic or multiple reproduction.

This research was supported by National Science Foundation grant MCS76-14294, and the Office of Naval Research grant number N00014-67-A-0204-0063.

* Note. This paper was submitted prior to the time that Rivest became editor of the department, and editorial consideration was completed under the former editor, G. K. Manacher.

Authors' Address: MIT Laboratory for Computer Science, 545 Technology Square, Cambridge, MA 02139.

© 1978 ACM 0001-0782/78/0200-0120 $00.75

120

I. Introduction

The era of "electronic mail" [10] may soon be upon us; we must ensure that two important properties of the current "paper mail" system are preserved: (a) messages are *private*, and (b) messages can be *signed*. We demonstrate in this paper how to build these capabilities into an electronic mail system.

At the heart of our proposal is a new encryption method. This method provides an implementation of a "public-key cryptosystem", an elegant concept invented by Diffie and Hellman [1]. Their article motivated our research, since they presented the concept but not any practical implementation of such a system. Readers familiar with [1] may wish to skip directly to Section V for a description of our method.

II. Public-Key Cryptosystems

In a "public-key cryptosystem" each user places in a public file an encryption procedure E. That is, the public file is a directory giving the encryption procedure of each user. The user keeps secret the details of his corresponding decryption procedure D. These procedures have the following four properties:

(a) Deciphering the enciphered form of a message M yields M. Formally,

D(E(M)) = M. (1)

(b) Both E and D are easy to compute.

(c) By publicly revealing E the user does not reveal an easy way to compute D. This means that in practice only he can decrypt messages encrypted with E, or compute D efficiently.

(d) If a message M is first deciphered and then enciphered, M is the result. Formally,

E(D(M)) = M. (2)

An encryption (or decryption) procedure typically consists of a *general method* and an *encryption key*. The general method, under control of the key, enciphers a message M to obtain the enciphered form of the message, called the *ciphertext* C. Everyone can use the same general method; the security of a given procedure will rest on the security of the key. Revealing an encryption algorithm then means revealing the key.

When the user reveals E he reveals a very *inefficient* method of computing D(C): testing all possible messages M until one such that E(M) = C is found. If property (c) is satisfied the number of such messages to test will be so large that this approach is impractical.

A function E satisfying (a)–(c) is a "trap-door one-way function;" if it also satisfies (d) it is a "trap-door one-way permutation." Diffie and Hellman [1] introduced the concept of trap-door one-way functions but

Communications of the ACM — February 1978, Volume 21, Number 2

Figure 2.6. The First Page of RSA's Paper

Example 2.2.1. The following functions are one-way functions:

(1) $f: pq \mapsto n$ is a one-way function, where p and q are prime numbers. The function f is easy to compute since the multiplication of p and q can be done in polynomial time. However, the computation of f^{-1}, the inverse of f is hard (this is the IFP problem).

(2) $f: x \mapsto g^x \bmod N$ is a one-way function. The function f is easy to compute since the modular exponentiation $g^x \bmod N$ can be performed in polynomial time. But the computation of f^{-1}, the inverse of f is hard (this is the DLP problem).

(3) $f: x \mapsto x^k \bmod N$ is a trapdoor one-way function, where $N = pq$ with p and q primes, and $kk' \equiv 1 \pmod{\phi(N)}$. It is obvious that f is easy

to compute since the modular exponentiation $x^k \bmod N$ can be done in polynomial time, but f^{-1}, the inverse of f (i.e., the kth root of x modulo N) is difficult to compute. However, if k', the trapdoor is given, f can be easily inverted, since $(x^k)^{k'} = x$.

Now we are in a position to introduce the formal definition of public-key cryptography.

Definition 2.2.2. A public-key cryptosystem $\mathcal{CS}$ may be formally defined as follows:

$$\mathcal{S} = (\mathcal{M}, \mathcal{C}, \mathcal{K}, M, C, e, d, E, D) \tag{2.7}$$

where

(1) $\mathcal{M}$ is the set of plaintexts, called the plaintext space.

(2) $\mathcal{C}$ is the set of cipherexts, called the ciphertext space.

(3) $\mathcal{K}$ is the set of keys, called the key space.

(4) $M \in \mathcal{M}$ is a piece of particular plaintext.

(5) $C \in \mathcal{C}$ is a piece of particular ciphertext.

(6) $e \neq d$ and $(e, d) \in \mathcal{K}$ is the key.

(7) E is the encryption function

$$E_{e_k} : M \mapsto C$$

where $M \in \mathcal{M}$ maps to $C \in \mathcal{C}$, using the public-key e_k, such that

$$C = E_{e_k}(M) \tag{2.8}$$

(8) D is the decryption function

$$D_{d_k} : C \mapsto M$$

where $C \in \mathcal{C}$ maps to $M \in \mathcal{M}$, using the private-key d_k such that

$$M = D_{d_k}(C) \tag{2.9}$$

satisfying

$$E_{e_k} D_{d_k} = 1 \text{ and } D_{d_k}(C) = D_{d_k}(E_{e_k}(M)) = M. \tag{2.10}$$

The main task in public-key cryptography is to find a suitable trap-door one-way function, so that both encryption and decryption are easy to perform for authorized users, whereas decryption, the inverse of the encryption, should be computationally infeasible for an unauthorized user.

2.3 RSA Public-Key Cryptography

This section introduces the basic idea and theory of the most popular and widely-used public-key cryptosystem RSA.

Definition 2.3.1. The *RSA public-key cryptosystem* may be formally defined as follows (Depicted in Figure 2.7):

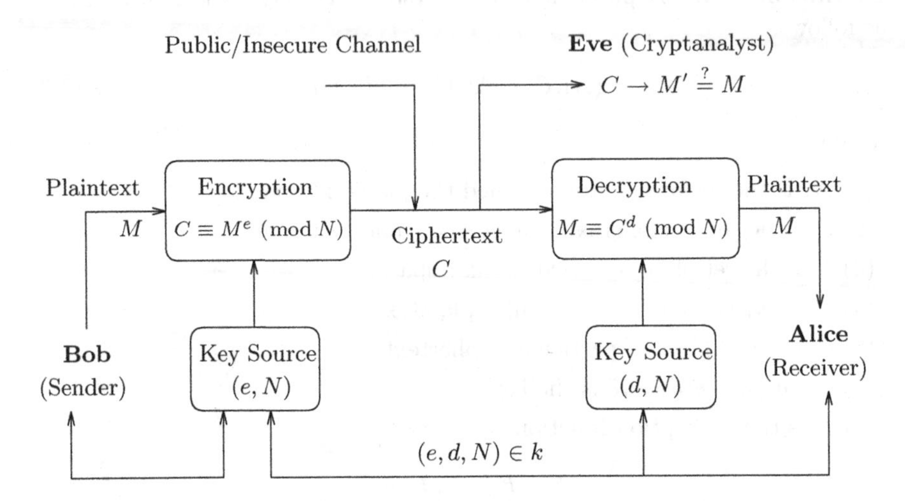

Figure 2.7. RSA Public-Key Cryptography

$$\text{RSA} = (\mathcal{M}, \mathcal{C}, \mathcal{K}, M, C, e, d, N, E, D) \tag{2.11}$$

where

(1) $\mathcal{M}$ is the set of plaintexts, called the plaintext space.

(2) $\mathcal{C}$ is the set of cipherexts, called the ciphertext space.

(3) $\mathcal{K}$ is the set of keys, called the key space.

(4) $M \in \mathcal{M}$ is a piece of particular plaintext.

(5) $C \in \mathcal{C}$ is a piece of particular ciphertext.

(6) $N = pq$ is the modulus with p, q prime numbers, usually each with at least 100 digits.

(7) $\{(e, N), (d, N)\} \in \mathcal{K}$ with $e \neq d$ are the encryption and encryption keys, respectively, satisfying

$$ed \equiv 1 \pmod{\phi(N)} \tag{2.12}$$

where $\phi(N) = (p-1)(q-1)$ is the Euler ϕ-function and defined by $\phi(N) = \#(\mathbb{Z}_N^*)$, the number of elements in the multiplicative group $\mathbb{Z}_N^*$.

(8) E is the encryption function

$$E_{e,N}: \; M \mapsto C$$

That is, $M \in \mathcal{M}$ maps to $C \in \mathcal{C}$, using the public-key (e, N), such that

$$C \equiv M^e \pmod{N}. \tag{2.13}$$

(9) D is the decryption function

$$D_{d,N}: \; C \mapsto M$$

That is, $C \in \mathcal{C}$ maps to $M \in \mathcal{M}$, using the private-key (d, N), such that

$$M \equiv C^d \equiv (M^e)^d \pmod{N}. \tag{2.14}$$

The idea of RSA can be best depicted in Figure 2.8.

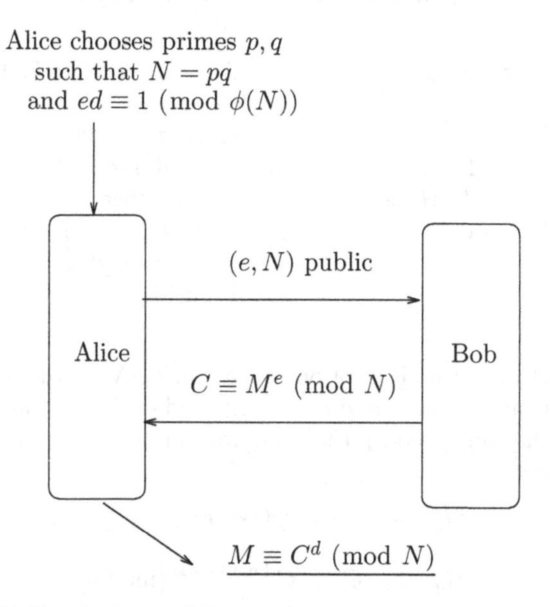

Figure 2.8. RSA Encryption and Decryption

Theorem 2.3.1 (The Correctness of RSA). Let M, C, N, e, d be plaintext, ciphertext, encryption exponent, decryption exponent, and modulus, respectively. Then

$$(M^e)^d \equiv M \pmod{N}.$$

Proof. Notice first that

$$\begin{aligned} C^d &\equiv (M^e)^d \pmod N && \text{(since } C \equiv M^e \pmod N\text{)} \\ &\equiv M^{1+k\phi(N)} \pmod N && \text{(since } ed \equiv 1 \pmod{\phi(N)}\text{)} \\ &\equiv M \cdot M^{k\phi(N)} \pmod N \\ &\equiv M \cdot (M^{\phi(N)})^k \pmod N \\ &\equiv M \cdot (1)^k \pmod N && \text{(by Euler's Theorem } a^{\phi(n)} \equiv 1 \pmod N\text{)} \\ &\equiv M \end{aligned}$$

The result thus follows. □

Both encryption $C \equiv M^e \pmod N$ and decryption $M \equiv C^d \pmod N$ of RSA can be implemented in polynomial-time by Algorithm 1.3.5. For example the RSA encryption can be implemented as follows:

Algorithm 2.3.1. Given (e, M, N), this algorithm finds $C \equiv M^e \pmod N$, or given (d, C, N), finds $M \equiv C^d \pmod N$ in time polynomial in $\log e$ or $\log d$, respectively.

```
Given (e, M, N) to find C
Set C <- 1
While e >= 1 do
        if e mod 2 = 1
          then C <- C · M mod N
        M <- M^2 mod N
        e <- ⌊e/2⌋
Print C
```

```
Given (d, C, N) to find M
Set M <- 1
While d >= 1 do
        if d mod 2 = 1
          then M <- M · C mod N
        C <- C^2 mod N
        d <- ⌊d/2⌋
Print M
```

Remark 2.3.1. For the decryption process in RSA, as the authorized user knows d and hence knows p and q, thus instead of directly working on $M \equiv C^d \pmod N$, he can speed-up the computation by working on the following two congruences:

$$\begin{aligned} M_p &\equiv C^d \equiv C^{d \bmod p-1} \pmod p \\ M_q &\equiv C^d \equiv C^{d \bmod q-1} \pmod q \end{aligned}$$

and then use the Chinese Remainder Theorem to get

$$M \equiv M_p \cdot q \cdot q^{-1} \bmod p + M_q \cdot p \cdot p^{-1} \bmod q \pmod N. \tag{2.15}$$

The Chinese Remainder Theorem is a two-edged sword. On the one hand, it provides a good way to speed-up the computation/performance of the RSA decryption, which can even be easily implemented by a low-cost crypto-chip [129]. On the other hand, it may introduce some serious security problems vulnerable to some side-channel attacks, particularly the random fault attacks; we shall discuss this in Section 8.4.

Example 2.3.1. Let the letter-digit encoding be as follows:

$$\text{space} = 00, A = 01, B = 02, \cdots, Z = 26.$$

(We will use this digital representation of letters throughout the book.) Let also

$$\begin{array}{rcl} e & = & 9007, \\ M & = & 2008050013010709030023151804190001180500191721050113 09\_ \\ & & 1908001519190906180107 05, \\ N & = & 114381625757888867669235779976146612010218296721242362\_ \\ & & 562561842935706935245733897830597123563958705058989075\_ \\ & & 147599290026879543541. \end{array}$$

Then the encryption can be done by using Algorithm 2.3.1:

$$\begin{array}{rcl} C & \equiv & M^e \\ & \equiv & 968696137546220614771409222543558829057599911245743198\_ \\ & & 746951209308162982251457083569314766228839896280133919\_ \\ & & 90551829945157815154 \pmod{N}. \end{array}$$

For the decryption, since the two prime factors p and q of N are known to the authorized person who does the decryption:

$$\begin{array}{rcl} p & = & 3490529510847650949147849619903898133417764638493387 8\_ \\ & & 43990820577, \\ q & = & 32769132993266709549961988190834461413177642967992942\_ \\ & & 539798288533, \end{array}$$

then

$$\begin{array}{rcl} d & \equiv & 1/e \\ & \equiv & 106698614368578024442868771328920154780709906633937862\_ \\ & \equiv & 801226224496631063125911774470873340168597462306553968\_ \\ & \equiv & 544513277109053606095 \pmod{(p-1)(q-1)}. \end{array}$$

Thus, the original plaintext M can be recovered either directly by using Algorithm 2.3.1, or indirectly by a combined use of Algorithm 2.3.1 and the Chinese Remainder Theorem (2.15):

$$\begin{array}{rcl} M & \equiv & C^d \\ & = & 200805001301070903002315180419000118050019172105011309\_ \\ & & 190800151919090618010705 \pmod{N} \end{array}$$

which is "THE MAGIC WORDS ARE SQUEAMISH OSSIFRAGE".

Remark 2.3.2. Prior to RSA, Pohlig and Hellman in 1978 [241] proposed a secret-key cryptography based on arithmetic modulo p, rather than $N = pq$. The Pohlig-Hellman system works as follows: Let M and C be the plain and cipher texts, respectively. Choose a prime p, usually with more than 200 digits, and a secret encryption key e such that $e \in \mathbb{Z}^+$ and $e \leq p-2$. Compute $d \equiv 1/e \pmod{(p-1)}$. (e, p) and of course d must be kept as a secret.

[1] **Encryption:**

$$C \equiv M^e \pmod{p}. \tag{2.16}$$

This process is easy for the authorized user:

$$\{M, e, p\} \xrightarrow[\text{easy}]{\text{find}} \{C \equiv M^e \pmod{p}\}. \tag{2.17}$$

[2] **Decryption:**

$$M \equiv C^d \pmod{p}. \tag{2.18}$$

For the authorized user who knows (e, p), this process is easy, since d can be easily computed from e.

[3] **Cryptanalysis:** The security of this system is based on the infeasibility of the Discrete Logarithm Problem. For example, for a cryptanalyst who does not know e or d would have to compute:

$$e \equiv \log_M C \pmod{p}.$$

Remark 2.3.3. One of the most important features of RSA encryption is that it can also be used for digital signatures. Let M be a document to be signed, and $N = pq$ with p, q primes, (e, d) the public and private exponents as in RSA encryption scheme. Then the processes of RSA signature signing and signature verification are just the same as that of the decryption and encryption; that is use d for signature signing and e signature verification as follows (see also Figure 2.9):

[1] **Signature signing**:

$$S \equiv M^d \pmod{N} \tag{2.19}$$

The signing process can only be done by the authorized person who has the private exponent d.

[2] **Signature verification**:

$$M \equiv S^e \pmod{N} \tag{2.20}$$

This verification process can be done by anyone since (e, N) is public.

Of course, RSA encryption and RSA signature can be used together to obtain a signed encrypted document to be sent over an insecure network.

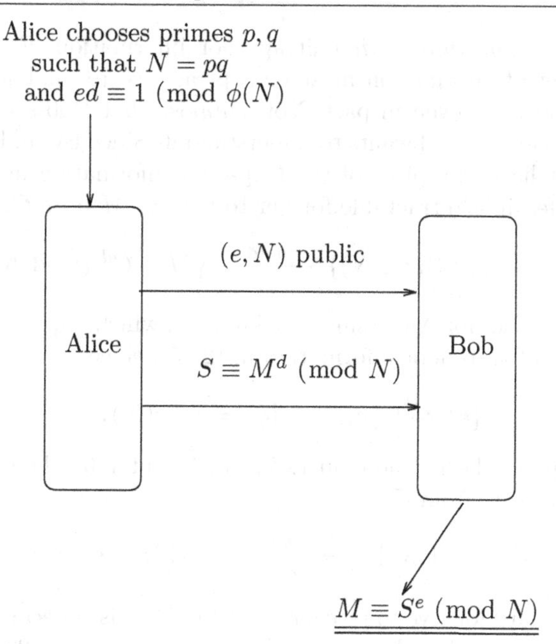

Figure 2.9. RSA Digital Signature

2.4 RSA Problem and RSA Assumption

As can be seen from the previous section, the whole idea of the RSA encryption and decryption is as follows:

$$\left.\begin{array}{rcl} C & \equiv & M^e \pmod N, \\ M & \equiv & C^d \pmod N \end{array}\right\} \tag{2.21}$$

where

$$\left.\begin{array}{rcl} ed & \equiv & 1 \pmod{\phi(N)} \\ N & = & pq \text{ with } p, q \in \text{ Primes.} \end{array}\right\} \tag{2.22}$$

Thus, the *RSA function* can be defined by

$$f_{\text{RSA}} : M \mapsto M^e \bmod N. \tag{2.23}$$

The *inverse of the RSA function* is then defined by

$$f_{\text{RSA}}^{-1} : M^e \mapsto M \bmod N. \tag{2.24}$$

Clearly, the RSA function is a *one-way trap-door function*, with

$$\{d, p, q, \phi(N)\} \tag{2.25}$$

the RSA *trap-door information*mitrap-door information. For security purposes, this set of information must be kept as a secret and should never be disclosed in anyway even in part. Now suppose that Bob sends C to Alive, but Eve intercepts it and wants to understand it. Since Eve only has (e, N, C) and does not have any piece of the trap-door information in (2.25), then it should be infeasible/intractable for her to recover M from C:

$$\{e, N, C \equiv M^e \pmod{N}\} \xrightarrow{\text{hard}} \{M \equiv C^d \pmod{N}\}. \tag{2.26}$$

On the other hand, for Alice, since she knows d, which implies that she knows all the pieces of trap-door information in (2.25), since

$$\{d\} \overset{\mathcal{P}}{\Longleftrightarrow} \{p\} \overset{\mathcal{P}}{\Longleftrightarrow} \{q\} \overset{\mathcal{P}}{\Longleftrightarrow} \{\phi(N)\} \tag{2.27}$$

We shall explain the relations in (2.27) in Chapter 6). Thus, it is easy for Alice to recover M from C:

$$\{N, C \equiv M^e \pmod{N}\} \xrightarrow[\text{easy}]{\{d,p,q,\phi(N)\}} \{M \equiv C^d \pmod{N}\}. \tag{2.28}$$

Why is it hard for Eve to recover M from C? This is because Eve is facing a hard computational problem, namely, the *RSA problem* [264]:

The RSA problem: Given the RSA public-key (e, N) and the RSA ciphertext C, find the corresponding RSA plaintext M. That is,

$$\{e, N, C\} \longrightarrow \{M\}.$$

It is conjectured although it has never been proved or disproved that:

The RSA conjecture: Given the RSA public-key (e, N) and the RSA ciphertext C, it is hard to find the corresponding RSA plaintext M. That is,

$$\{e, N, C\} \xrightarrow{\text{hard}} \{M\}.$$

But how hard is it for Alice to recover M from C? This is another version of the RSA conjecture, often called the *RSA assumption*, which again has never been proved or disproved:

The RSA assumption: Given the RSA public-key (e, N) and the RSA ciphertext C, then finding M is as hard as factoring the RSA modulus N. That is,

$$\text{IFP}(N) \Longleftrightarrow \text{RSA}(M)$$

provided that N is sufficiently large and randomly generated, and M and C are random integers between 0 and $N-1$. More precisely, it is conjectured (or assumed) that

$$\text{IFP}(N) \overset{\mathcal{P}}{\Longleftrightarrow} \text{RSA}(M).$$

That is, if N can be factorized in polynomial-time, then M can be recovered from C in polynomial-time. In other words, cryptoanalyzing RSA must be as difficult as solving the IFP problem. But the problem is, as we discussed previously, that no one knows whether or not IFP can be solved in polynomial-time, so RSA is only assumed to be secure, not proved to be secure:

$$\text{IFP}(N) \text{ is hard } \longrightarrow \text{RSA}(M) \text{ is secure.}$$

The real situtaion is that

$$\text{IFP}(N) \overset{\surd}{\Longrightarrow} \text{RSA}(M),$$

$$\text{IFP}(N) \overset{?}{\Longleftarrow} \text{RSA}(M).$$

Now we can return to answer the question that how hard is it for Alice to recover M from C? By the RSA assumpition, cryptanalyzing C is as hard as factoring N. The fastest known integer factorization algorithm, the Number Field Sieve (NFS), runs in time

$$\mathcal{O}(\exp(c(\log N)^{1/3}(\log\log N)^{2/3}))$$

where $c = (64/9)^{1/3}$ if a general version of NFS, GNFS, is used for factoring an arbitrary integer N whereas $c = (32/9)^{1/3}$ if a special version of NFS, SNFS, is used for factoring a special form of integer N. As in RSA, the modululs $N = pq$ is often chosen be a large general composite integer $N = pq$ with p and q the same bit size, which makes SNFS is not useful. This means that RSA cannot be broken in polynomial-time, but in subexponential-time, which makes RSA secure, again, by assumption. Thus, readers should note that the RSA problem is *assumed* to be *hard*, and the RSA cryptosystem is *conjectured* to be *secure* .

2.5 RSA-Type Crytposystems

RSA is a cryptographic system based on factoring as its security relies on the intractability of the Integer Factorization Problem (IFP). However, RSA is not the only cryptographic system based on factoring. There are in fact many cryptographic systems whose security depends on the intractability of the IFP problem. In this section, we study some of these systems. If the RSA problem can be solved in polynomial-time, all the factoring based cryptographic systems can be broken in polynomial-time. Thus, we could regard all the factoring based cryptographic systems as various variants of the RSA cryptographic systems, or *RSA-type cryptosystems.* In a more broad sense, as IFP is related to DLP/ECDLP, all DLP/ECDLP-based cryptosystems may also be regarded as RSA-type cryptosystems.

Rabin's M^2 Cryptoystem

As can be seen from the previous sections, RSA uses M^e for encryption, with $e \geq 3$ (3 is the smallest possible public exponent), we might call RSA encryption M^e encryption. In 1979, Michael Rabin proposed a scheme based on M^2 encryption. rather than the M^e for $e \geq 3$ encryption used in RSA. A brief description of the Rabin system is as follows (see also Figure 2.10).

[1] **Key generation:** Let $N = pq$ with p, q odd primes satisfying

$$p \equiv q \equiv 3 \pmod 4. \tag{2.29}$$

[2] **Encryption:**

$$C \equiv M^2 \pmod N. \tag{2.30}$$

[3] **Decryption:** Use the Chinese Remainder Theorem to solve the system of congruences:

$$\begin{cases} M_p \equiv \sqrt{C} \pmod p \\ M_q \equiv \sqrt{C} \pmod q \end{cases} \tag{2.31}$$

to get the four solutions: $\{\pm M_p, \pm M_q\}$. The true plaintext M will be one of these four values.

[4] **Cryptanalysis:** A cryptanalyst who can factor N can compute the four square roots of C modulo N, and hence can recover M from C. Thus, breaking the Rabin system is equivalent to factoring N.

Unlike the RSA cryptosystem whose security was only conjectured to be equivalent to the intractability of IFP, the security of Rabin-Williams is proved to be equivalent to the intractability of IFP. First notice that there is a fast algorithm to compute the square roots modulo N if $N = pq$ is known.

Consider the following quadratic congruence

$$x^2 \equiv y \pmod p \tag{2.32}$$

there are essentially three cases for the prime p:

(1) $p \equiv 3 \pmod 4$,
(2) $p \equiv 5 \pmod 8$,
(3) $p \equiv 1 \pmod 8$.

All three cases may be solved by the following process:

$$\begin{cases} \text{if } p \equiv 3 \pmod 4, & x \equiv \pm y^{\frac{p+1}{4}} \pmod p, \\ \text{if } p \equiv 5 \pmod 8, & \begin{cases} \text{if } y^{\frac{p+1}{4}} = 1, & x \equiv \pm y^{\frac{p+3}{8}} \pmod p \\ \text{if } y^{\frac{p+1}{4}} \neq 1, & x \equiv \pm 2y(4y)^{\frac{p-5}{8}} \pmod p. \end{cases} \end{cases} \tag{2.33}$$

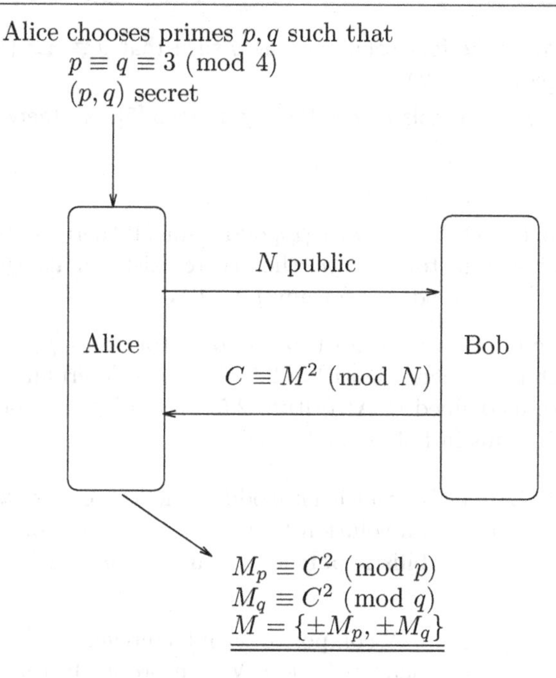

Figure 2.10. Rabin System

Algorithm 2.5.1 (Computing Square Roots Modulo pq). Let $N = pq$ with p and q odd prime and $y \in \mathrm{QR}_N$. This algorithm will find all the four solutions in x to congruence $x^2 \equiv y \pmod{pq}$ in time $\mathcal{O}((\log p)^4)$.

[1] Use (2.33) to find a solution r to $x^2 \equiv y \pmod p$.

[2] Use (2.33) to find a solution s to $x^2 \equiv y \pmod q$.

[3] Use the Extended Euclid's algorithm to find integers c and d such that $cp + dq = 1$.

[4] Compute $x \equiv \pm(rdq \pm scp) \pmod{pq}$.

On the other hand, if there exists an algorithm to find the four solutions in x to $x^2 \equiv y \pmod N$, then there exists an algorithm to find the prime factorization of N. The following is the algorithm.

Algorithm 2.5.2 (Factoring via Square Roots). This algorithm seeks to find a factor of N by using an existing square root finding algorithm (namely, Algorithm 2.5.1).

[1] Choose at random an integer x such that $\gcd(x, N) = 1$, and compute $x^2 \equiv a \pmod N$.

[2] Use Algorithm 2.5.1 to find four solutions in x to $x^2 \equiv a \pmod N$.

[3] Choose one of the four solutions, say y such that $y \not\equiv \pm x \pmod N$, then compute $\gcd(x \pm y, N)$.

[4] If $\gcd(x \pm y, N)$ reveals p or q, then go to step [5], or otherwise, go to step [1].

[5] Exit.

Theorem 2.5.1. Let $N = pq$ with p, q odd prime. If there exists a polynomial-time algorithm A to factor $N = pq$, then there exists an algorithm B to find a solution to $x^2 \equiv y \pmod N$, for any $y \in \mathrm{QR}_N$.

Proof. If there exists an algorithm A to factor $N = pq$, then there exists an algorithm (in fact, Algorithm 2.5.1), which determines $x = \pm(rdq \pm scp) \pmod{pq}$, as defined in Algorithm 2.5.1, for $x^2 \equiv y \pmod N$. Clearly, Algorithm 2.5.1 runs in polynomial-time. □

Theorem 2.5.2. Let $N = pq$ with p, q odd prime. If there exists a polynomial-time algorithm A to find a solution to $x^2 \equiv a \pmod N$, for any $a \in \mathrm{QR}_N$, then there exists a probabilistic polynomial time algorithm B to find a factor of N.

Proof. First note that for N composite, x and y integer, if $x^2 \equiv y^2 \pmod N$ but $x \not\equiv \pm y \pmod N$, then $\gcd(x + y, N)$ are proper factors of n. If there exists an algorithm A to find a solution to $x^2 \equiv a \pmod N$ for any $a \in \mathrm{QR}_N$, then there exists an algorithm (in fact, Algorithm 2.5.2), which uses algorithm A to find four solutions in x to $x^2 \equiv a \pmod N$ for a random x with $\gcd(x, N) = 1$. Select one of the solutions, say, $y \not\equiv \pm x \pmod N$, then by computing $\gcd(x \pm y, N)$, the probability of finding a factor of N will be $\geq 1/2$. If Algorithm 2.5.2 runs for k times and each time randomly chooses a different x, then the probability of not factoring N is $\leq 1/2^k$. □

So, finally, we have

Theorem 2.5.3. Factoring integers, computing the modular square roots, and breaking the Rabin cryptosystem are computationally (deterministic polynomial-time) equivalent. That is,

$$\mathrm{IFP}(N) \overset{\mathcal{P}}{\Longleftrightarrow} \mathrm{Rabin}(M). \tag{2.34}$$

Williams' Improved M^2 Cryptoystem

Williams [327] proposed a modified version of the RSA cryptographic system, particularly the Rabin's M^2 system in order to make it suitable as a public-key encryption scheme (Rabin's original system was intended to be used as a digital signature scheme). A description of *Williams' M^2 encryption* is as follows (suppose Bob wishes to send Alice a ciphertext $C \equiv M^2 \pmod N$):

[1] **Key generation:** Let $N = pq$ with q and q primes such that

$$\left.\begin{array}{l} p \equiv 3 \pmod 8 \\ q \equiv 7 \pmod 8 \end{array}\right\}$$

So, $N \equiv 5 \pmod 8$ and $(N, 2)$ is used public-key. The private-key d is defined by

$$d = \frac{(p-1)(q-1)}{4} + 1$$

[2] **Encryption:** Let $\mathcal{M}$ be plaintext space containing all possible plaintexts M such that

$$2(2M+1) < N \text{ if the Jacobi symbol } \left(\frac{2M+1}{N}\right) = -1$$

$$4(2M+1) < N \text{ if the Jacobi symbol } \left(\frac{2M+1}{N}\right) = 1$$

The first step in encryption is for all $M \in \mathcal{M}$, put

$$M' = E_1(M) = \begin{cases} 2(2M+1) \text{ if the Jacobi symbol } \left(\frac{2M+1}{N}\right) = -1 \\ 4(2M+1) \text{ if the Jacobi symbol } \left(\frac{2M+1}{N}\right) = 1. \end{cases}$$

The last step in encryption is just the same as Rabin's encryption:

$$C \equiv (M')^2 \pmod N$$

[3] **Decryption:** On the reverse order of the encryption, the first step in decryption is as follows:

$$C' = D_2(C) \equiv C^d \pmod N$$

and the last step in decryption is defined by:

$$M = D_1(C') = \begin{cases} \frac{\frac{M'}{4} - 1}{2} & \text{if } M' \equiv 0 \pmod 4 \\ \frac{\frac{N-M'}{4} - 1}{2} & \text{if } M' \equiv 1 \pmod 4 \\ \frac{\frac{M'}{2} - 1}{2} & \text{if } M' \equiv 2 \pmod 4 \\ \frac{\frac{N-M'}{2} - 1}{2} & \text{if } M' \equiv 3 \pmod 4. \end{cases}$$

The whole process of encryption and decryption is as follows:

$$M \xrightarrow{E_1} M' \xrightarrow{E_2} C \xrightarrow{D_2} M' \xrightarrow{D_1} M.$$

[4] **Cryptanalysis:** A cryptanalyst who can factor N can find d, and hence can recover M from C. Thus, breaking the Williams' system is equivalent to factoring N.

Theorem 2.5.4 (Correctness of Williams' M^2 encryption). Let $M \in \mathcal{M}$. Then

$$M = D_1(D_2(E_2(E_1(M)))).$$

Theorem 2.5.5 (Equivalence of Williams(M) and IFP(N)). Breaking Williams' M^2 encryption (i.e., finding M from C) is equivalent to factoring the modulus N. That is,

$$\text{IFP(N)} \overset{\mathcal{P}}{\Longleftrightarrow} \text{Williams}(M). \tag{2.35}$$

For the justification od the above two theorem, see Williams [327].

Just the same as Rabin's system, Williams' M^2 encryption is also provably secure, as breaking the Williams' M^2 mod N encryption is equivalent to factoring N, where the N is a special form of $N = pq$, with p, q primes and $p \equiv 3 \pmod 8$ and $q \equiv 7 \pmod 8$. Note that this special integer factorization problem is not the same as the general IFP problem, although there is no any known reason to believe this special factoring problem is any easier than the general factoring problem. But unlike Rabin's system, Williams' M^2 encryption can be easily generalized to the general M^e encryption with $e > 2$, as in RSA. Thus, Williams' M^2 encryption is not just a variant of Rabin system, but also a variant of the general RSA system. In fact, Williams' original paper [327] discussed the general case that

$$ed \equiv \frac{\frac{(p-1)(q-1)}{4} + 1}{2} \pmod{\lambda(N)}.$$

Williams' M^2 encryption improved Rabin's M^2 encryption by eliminating the 4 : 1 ciphertext ambiguity problem in decryption without adding extra information for removing the ambiguity. Williams in [331] also proposed a M^3 *encryption* variant to Rabin but eliminated the 9 : 1 ciphertext ambiguity problem. The encryption is also proved to be as hard as factoring, although is is again still not the general IFP problem, since $N = pq$ was chosen to be

$$p \equiv q \equiv 1 \pmod 3$$

and

$$\frac{(p-1)(q-1)}{9} \equiv -1 \pmod 3.$$

LUC Cryptoystem

In 1993, Smith and Lennon proposed an RSA analog, called LUC [305], based on Lucas sequences. Let a, b be non-zero integers and $D = a^2 - 4b$. Consider the equation $x^2 - ax + b = 0$; its discriminant is $D = a^2 - 4b$, and α and β are the two roots:

$$\begin{aligned} \alpha &= \frac{a + \sqrt{D}}{2} \\ \beta &= \frac{a - \sqrt{D}}{2}. \end{aligned}$$

So

$$\begin{aligned} \alpha + \beta &= a \\ \alpha - \beta &= \sqrt{D} \\ \alpha\beta &= b. \end{aligned}$$

We define the sequences (U_k) and (V_k) by

$$\begin{aligned} U_k(a, b) &= \frac{\alpha^k - \beta^k}{\alpha - \beta} \\ V_k(a, b) &= \alpha^k + \beta^k. \end{aligned}$$

In particular, $U_0(a, b) = 0, U_1(a, b) = 1$, while $V_0(a, b) = 2, V_1(a, b) = a$. For $k \geq 2$, we also have

$$\begin{aligned} U_k(a, b) &= aU_{k-1} - bU_{k-2} \\ V_k(a, b) &= aV_{k-1} - bV_{k-2}. \end{aligned}$$

The sequences

$$\begin{aligned} U(a, b) &= (U_k(a, b))_{k \geq 0} \\ V(a, b) &= (V_k(a, b))_{k \geq 0} \end{aligned}$$

are called the *Lucas sequences* associated with the pair (a, b), in honour of the French mathematician François Edouard Lucas (1842–1891); more information about Lucas and Lucas sequences can be found e.g., in Yan [335]. The LUC cryptosystem works as follows (suppose Bob sends a ciphertext to Alice):

[1] **Key generation:** Alice publishes her public-key (e, N), satisfying

$$\begin{aligned} N = pq, &\quad \text{with} \quad p, q \in \text{Primes}, \\ ed \equiv 1 \pmod{(p^2 - 1)(q^2 - 1)}, &\quad \text{with} \quad \gcd(e, (p^2 - 1)(q^2 - 1)) = 1. \end{aligned}$$

[2] **Encryption:** Bob encrypts his message $1 < M < N-1$ with $\gcd(M, N) = 1$ as follows:

$$C \equiv V_e(M, 1) \pmod{N}$$

based on the Lucas sequences, and sends it to Alice.

[3] **Decryption:** Alice performs the decryption as follows:

$$M \equiv v_d(C, 1) \pmod{N}.$$

[4] **Cryptoanalysis:** Anyone who can factor N can decrypt the message.

Theorem 2.5.6 (Correctness of LUC).

$$v_d(C, 1) \equiv M \pmod{N}.$$

Proof.

$$\begin{aligned} v_d(C, 1) &\equiv v_d(V_e(M, 1), 1) \\ &\equiv v_{ed}(M, 1) \\ &\equiv M \pmod{N}. \end{aligned}$$

□

Readers are suggested to consult [33], [205], [237] and [305] for more information about the LUC system, [33] also had a good discussion on both the LUC and the Dickson cryptosystems, as Dickson polynomials and the Lucas sequences are related to each other, and both can be used to construct RSA-type cryptosystems.

Elliptic Curve RSA

RSA has several noted elliptic curve analogues. Before introducing ing the EC analogue of RSA, we need two more results related the number of points on elliptic curves over the finite field $\mathbb{F}_p$.

Theorem 2.5.7 (Hasse). Let E be an elliptic curve over $\mathbb{F}_p$:

$$E_p(a, b) : \; y^2 = x^3 + ax + b. \tag{2.36}$$

Then the number of points on $E_p(a, b)$, denoted by $\#(E_p(a, b))$, is as follows:

$$\#((E_p(a, b)) = 1 + p - \epsilon$$

where $|\epsilon| \leq 2\sqrt{p}$.

Definition 2.5.1. Let $E_p(a, b)$ be an elliptic curve over the finite field $\mathbb{F}_p$. The complementary group of $E_p(a, b)$, denoted by $\overline{E_q(a, b))}$, is the set of points satisfying (2.36) together with a point at infinity $\mathcal{O}$, where y is of the form $u\sqrt{v}$, and v is a fixed non-quadratic residue modulo p and $v \in \mathbb{F}_p$.

Corollary 2.5.1. If

$$\#((E_p(a,b)) = 1 + p - \epsilon,$$

then

$$\#(\overline{E_q(a,b)})) = 1 + p + \epsilon.$$

The simplest EC analogue of RSA may be described as follows [335]:

[1] **Key generation:** $N = pq$ is the product of two large secret primes p and q. Choose two random integers a and b such that $E_N(a,b): y^2 = x^3 + ax + b$ defines an elliptic curve modulo both p and q. Let

$$N_n = \text{lcm}(\#(E_p(a,b)), \#(E_q(a,b))).$$

Choose a value for e such that

$$\begin{aligned} \gcd(e, N_n) &= 1, \\ ed &\equiv 1 \pmod{N_n}. \end{aligned}$$

Publish (e, N) as public-key, but keep (d, p, q, N_n) as a secret.

[2] **Encryption:** To encrypt a message-point M, which is a point on $E_N(a,b)$, perform $C \equiv eM \pmod{N}$.

[3] **Decryption:** To encrypt a ciphertext C, just perform $M \equiv dC \pmod{N}$.

[4] **Cryptanalysis:** Anyone who can factor N, can of course decode the ciphertext.

In what follows, we introduce two relatively popular elliptic curve analogues of RSA. The first is the *KMOV cryptosystem* [180], which uses a family of supersingular elliptic curve $E_N(0,b): y^2 = x^3 + b$. An important property of this system is that if

$$p, q \equiv 2 \pmod{3},$$

then

$$N_n = \text{lcm}(p+1, q+1)$$

regardless of the value of b. The KMOV systems works as follows:

[1] **Key generation:** Let $N = pq$ be the product of two large secret primes p and q. Choose a supersingular elliptic curve

$$E_N(0,b): y^2 = x^3 + b$$

such that

$$N_n = \text{lcm}(p+1, q+1).$$

Choose a value for e such that

$$\begin{aligned} \gcd(e, N_n) &= 1, \\ ed &\equiv 1 \pmod{N_n}. \end{aligned}$$

Publish (e, N) as public-key, but keep (d, p, q, N_n) as a secret.

[2] **Encryption:** To encrypt a message, e.g., $M = (m_1, m_2)$, Choose a suitable $b$such that

$$b \equiv m_2^3 - m_1^3 \pmod N$$

and C is computed by

$$C \equiv eM \pmod N$$

over $E_N(0, b)$.

[3] **Decryption:** To encrypt a ciphertext C, perform

$$M \equiv dC \pmod N$$

over $E_N(0, b)$.

[4] **Cryptanalysis:** Anyone who can factor N, can find the trap-door information (d, p, q, N_n), and hence can decode the message.

The second is the Demytko cryptosystem [98], which uses fixed (a, b) for elliptic curve

$$E_N(a, b): \ y^2 = x^3 + ax + b,$$

In particular, it uses only the x-coordinate of the points of $E_p(a, b)$, The system relies on the fact if x is not the x-coordinate of a point on the elliptic curve $E_N(a, b)$, then it will be the x-coordinate of a point on of the *twisted* curve $\overline{E_p(a, b)}$. Thus,

$$N_n = \text{lcm}(\#(E_p(a, b)), \#(\overline{E_p(a, b)}), \#(E_q(a, b)), \#(\overline{E_q(a, b)})).$$

The Demytko system works as follows:

[1] **Key generation:** Choose $N = pq$ with p and q primes. Choose an elliptic curve

$$E_N(0, b): \ y^2 = x^3 + b$$

with fixed parameters p and q. Let

$$N_n = \text{lcm}(\#(E_p(a, b)), \#(\overline{E_p(a, b)}), \#(E_q(a, b)), \#(\overline{E_q(a, b)})).$$

Choose a value for e such that

$$\begin{aligned} \gcd(e, N_n) &= 1, \\ ed &\equiv 1 \pmod{N_n}. \end{aligned}$$

Publish (e, N) as public-key, but keep (d, p, q, N_n) as a secret.

[2] **Encryption:** To encrypt a message M, compute

$$C \equiv eM \pmod N$$

over $E_N(a, b)$.

[3] **Decryption:** To encrypt a ciphertext C, perform

$$M \equiv dC \pmod{N}$$

over $E_N(a, b)$.

[4] **Cryptanalysis:** Anyone who can factor N, can recover M from C.

The cryptanalyst who knows the prime factorization of N, can decrypt the message easily. Readers are advised to consult [158] and [159] for some more recent developments in elliptic curve analogues of RSA.

ElGamal System

RSA is not only connected to the integer factorization problem, but also connected to the discrete logarithm problem, since, e.g., M can be found by taking the following discrete logarithm:

$$M \equiv \log_{M^e} M \pmod{N}$$

The first public-key cryptosystem based on discrete logarithms is the ElGamal cryptosystem (see Figure 2.11), proposed in 1985:

[1] A prime q and a generator $g \in \mathbb{F}_q^*$ are made public.

[2] Alice chooses a private integer $a = a_A \in \{1, 2, \cdots, q-1\}$. This a is the private decryption key. The public encryption key is $g^a \in \mathbb{F}_q$.

[3] Suppose now Bob wishes to send a message to Alice. He chooses a random number $b \in \{1, 2, \cdots, q-1\}$ and sends Alice the following pair of elements of $\mathbb{F}_q$:

$$(g^b, \; Mg^{ab})$$

where M is the message.

[4] Since Alice knows the private decryption key a, she can recover M from this pair by computing $g^{ab} \pmod{q}$ and dividing this result into the second element, i.e., Mg^{ab}.

[5] Someone who can solve the discrete logarithm problem in $\mathbb{F}_q$ breaks the cryptosystem by finding the secret decryption key a from the public encryption key g^a. In theory, there could be a way to use knowledge of g^a and g^b to find g^{ab} and hence break the cipher without solving the discrete logarithm problem. However, there is no known way to go from g^a and g^b to g^{ab} without essentially solving the discrete logarithm problem. So the security of the ElGamal cryptosystem is the same as the intractability of the discrete logarithm problem.

Surprisingly, the ElGamal cryptosystem (and in fact, almost all the existing systems, including RSA) can be easily extended to an elliptic curve cryptosystem. The following is the elliptic curve analog of the ElGamal cryptosystem (see also Figure 2.12).

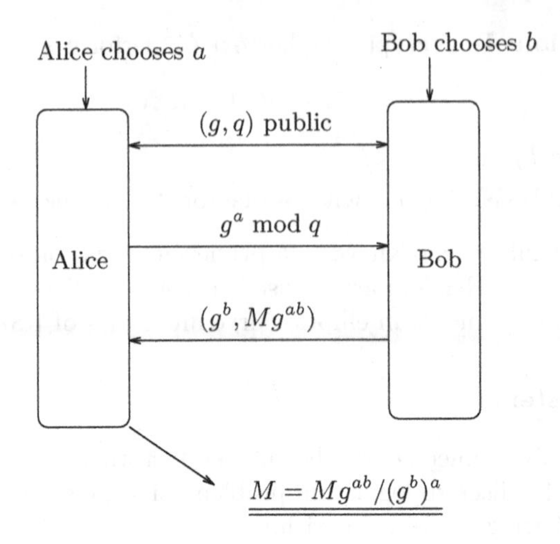

Figure 2.11. ElGamal System

[1] Alice and Bob publicly choose an elliptic curve E over $\mathbb{F}_q$ with $q = p^r$ and $p \in$ Primes, and a random *base* point $P \in E$.

[2] Alice chooses a random integer r_a and computes r_aP; Bob also chooses a random integer r_b and computes r_bP.

[3] To send a message-point M to Bob, Alice chooses a random integer k and sends the pair of points $(kP,\ M + k(r_bP))$.

[4] To read M, Bob computes

$$M + k(r_bP) - r_b(kP) = M. \tag{2.37}$$

[5] An eavesdropper who can solve the discrete logarithm problem on E can, of course, determine r_b from the publicly known information P and r_bP. Since there is no known efficient way to compute discrete logarithms, the system is secure.

Goldwasser-Micali System

The RSA encryption is *deterministic* in the sense that under a fixed public-key, a particular plaintext M is always encrypted to the same ciphertext C. Some of the drawbacks of such a deterministic scheme are:

(1) It is not secure for all probability distributions of the message space. For example, in RSA encryption, the messages 0 and 1 always get encrypted to themselves, and hence are easy to detect.

(2) It is easy to obtain some partial information of the secret key (p, q) from the public modulus n (assume that $n = pq$). For example, when the least-significant digit of n is 3, then it is easy to obtain the partial

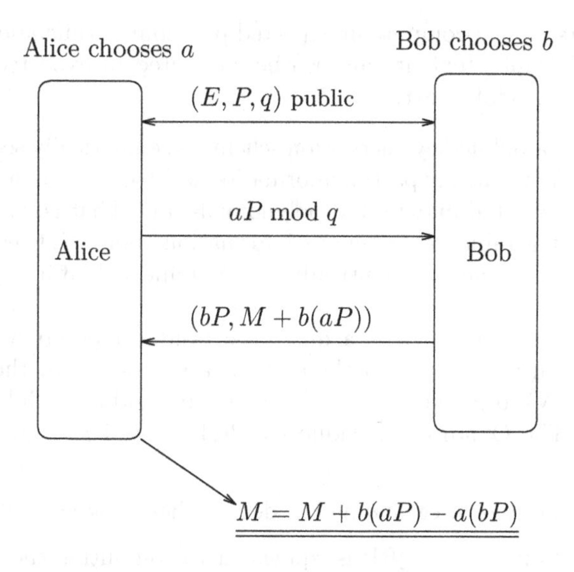

Figure 2.12. Elliptic Curve Analog of ElGamal System

information that the least-significant digits of p and q are either 1 and 3 or 7 and 9 since

$$\begin{array}{ll} 183 = 3 \cdot 61 & 253 = 11 \cdot 23 \\ 203 = 7 \cdot 29 & 303 = 3 \cdot 101 \\ 213 = 3 \cdot 71 & 323 = 17 \cdot 19. \end{array}$$

(3) It is sometimes easy to compute partial information about the plaintext M from the ciphertext C. For example, given (C, e, N), the Jacobi symbol of M over N can be easily deduced from C:

$$\left(\frac{C}{N}\right) = \left(\frac{M^e}{N}\right) \left(\frac{M}{N}\right)^e = \left(\frac{M}{N}\right).$$

d) It is easy to detect when the same message is sent twice.

Probabilistic encryption, or randomized encryption, however, utilizes randomness to attain a strong level of security, namely, the *polynomial security* and *semantic security*, defined as follows:

Definition 2.5.2. A public-key encryption scheme is said to be *polynomially* secure if no passive adversary can, in expected polynomial-time, select two plaintexts M_1 and M_2 and then correctly distinguish between encryptions of M_1 and M_2 with probability significantly greater that $1/2$.

Definition 2.5.3. A public-key encryption scheme is said to be *semantically* secure if, for all probability distributions over the message space, whatever a passive adversary can compute in expected polynomial-time about the plaintext given the ciphertext, it can also be computed in expected polynomial time without the ciphertext.

Intuitively, a public-key encryption scheme is semantically secure if the ciphertext does not leak any partial information whatsoever about the plaintext that can be computed in expected polynomial-time. That is, given (C, e, N), it should be intractable to recover any information about M. Clearly, a public-key encryption scheme is semantically secure if and only if it is polynomially secure.

Recall that an integer a is a quadratic residue modulo N, denoted by $a \in Q_N$, if $\gcd(a, N) = 1$ and there exists a solution x to the congruence $x^2 \equiv a \pmod{N}$, otherwise a is a quadratic nonresidue modulo N, denoted by $a \in \overline{Q}_N$. The Quadratic Residuosity Problem (QRP) may be stated as follows:

Given positive integers a and n, decide whether or not $a \in Q_N$.

It is believed that solving QRP is equivalent to computing the prime factorization of N, so it is computationally infeasible. The Jacobi symbol $\left(\frac{x}{N}\right)$ is defined for any $x \in \mathbb{Z}_N$ and has a value in $\{1, -1\}$. If N is prime then

$$a \in Q_N \iff \left(\frac{a}{N}\right) = 1 \tag{2.38}$$

and if N is composite, then

$$a \in Q_N \implies \left(\frac{a}{N}\right) = 1 \tag{2.39}$$

but

$$a \in Q_N \stackrel{?}{\Longleftarrow} \left(\frac{a}{N}\right) = 1. \tag{2.40}$$

However

$$a \in \overline{Q}_N \Longleftarrow \left(\frac{a}{N}\right) = -1. \tag{2.41}$$

That is, whenever N is composite, a may belong to $\overline{Q}_N$ even if $\left(\frac{a}{N}\right) = 1$. Let $J_n = \{a \in (\mathbb{Z}/n\mathbb{Z})^* : \left(\frac{a}{N}\right) = 1\}$, then $\tilde{Q}_N = J_N - Q_N$. Thus, $\tilde{Q}_N$ is the set of all pseudosquares modulo N; it contains those elements of J_N that do not belong to Q_N.

In what follows, we present a cryptosystem whose security is based on the infeasibility of the Quadratic Residuosity Problem; it was first proposed by Goldwasser and Micali in 1984, under the term *probabilistic encryption.*

Algorithm 2.5.3 (Goldwasser-Micali Probabilistic Encryption). This algorithm uses the randomized method to encrypt messages and is based on the quadratic residuosity problem (QRP). The algorithm divides into three parts: key generation, message encryption and decryption.

[1] **Key generation**: Both Alice and Bob should do the following to generate their public and secret keys:
  - Select two large distinct primes p and q, each with roughly the same size, say, each with β bits.
  - Compute $N = pq$.
  - Select a $y \in \mathbb{Z}_N$, such that $y \in \overline{Q}_N$ and $\left(\frac{y}{N}\right) = 1$. ($y$ is thus a pseudosquare modulo N).
  - Make (N, y) public, but keep (p, q) secret.

[2] **Encryption**: To send a message to Alice, Bob should do the following:
  - Obtain Alice's public-key (N, y).
  - Represent the message m as a binary string $m = m_1m_2 \cdots m_k$ of length k.
  - For i from 1 to k do
    - Choose at random an $x \in (\mathbb{Z}_N)^*$ and call it x_i.
    - Compute c_i:

$$c_i = \begin{cases} x_i^2 \bmod N, & \text{if } m_i = 0, \quad \text{(r.s.)} \\ yx_i^2 \bmod N, & \text{if } m_i = 1, \quad \text{(r.p.s.)}, \end{cases} \tag{2.42}$$

    where r.s. and r.p.s. represent random square and random pseudosquare, respectively.
    - Send the k-tuple $c = (c_1, c_2, \cdots, c_k)$ to Alice. (Note first that each c_i is an integer with $1 \le c_i < N$. Note also that since n is a 2β-bit integer, it is clear that the ciphertext c is a much longer string than the original plaintext m.)

[3] **Decryption**: To decrypt Bob's message, Alice should do the following:
  - For i from 1 to k do
    - Evaluate the Legendre symbol:

$$e_i' = \left(\frac{c_i}{p}\right)$$

    - Compute m_i:

$$m_i = \begin{cases} 0, & \text{if } e_i' = 1 \\ 1, & \text{if otherwise.} \end{cases} \tag{2.43}$$

    That is, $m_i = 0$ if $c_i \in Q_N$, otherwise, $m_i = 1$.
  - Finally, get the decrypted message $m = m_1m_2 \cdots m_k$.

One of the most important feature of the Goldwasser-Micali encryption is that

Theorem 2.5.8. The Goldwasser-Micali probabilistic encryption based on QRP is semantically secure.

Readers interested in semantically secure probabilistic encryption may wish to consult [124] and [121].

2.6 Chapter Notes and Further Readings

The underlying number-theoretic and computational complexity-theoretic ideas and concepts of the public-key cryptography in general and RSA cryptography in particular have been introduced in this chapter.

The idea of public-key cryptography was first publicly conceived and proposed by Diffie and Hellman [101], then both at Stanford University, in their seminal paper "New Directions in Cryptography" [101]; which was in turn based on some idea of Merkle [208], a PhD student of Hellman. Based on the work of Diffie, Hellman and Merkle (DHM), Rivest, Shamir and Addleman (RSA), then all at MIT, developed the first practical public-key cryptosystem in 1977 (see [115], [261], and [262]), now known as the RSA cryptosystem. These six people Diffie, Hellman, Merkle, Rivest, Shamir and Addleman are now regarded as the co-inventors of the public-key cryptography. Incidently, they jointly received the 1996 ACM Paris Kanellakis Theory and Practice Award "for the conception and first effective realization of public-key cryptography. The idea of a public-key cryptosystem was a major conceptual breakthrough that continues to stimulate research to this day, and without it today's rapid growth of electronic commerce would have been impossible." It is interesting to note that in December 1997 the Communication-Electronics Security Group (CESG) of the British Government Communications Headquarters (GCHQ), the successor of the Blechley Park (the British Government's Cryptography Centre during the 2nd World War for breaking the German Enigma codes), claimed that public-key cryptography was conceived (secretly) by James H. Ellis in 1970 and implemented by two of his colleagues Clifford C. Cocks and Malcolm J. Williamson between 1973 and 1976 in CESG, by declassifying five of their papers. The US Government's National Security Agency (NSA) also made a similar claim that they had public-key cryptography a decade earlier. However, according to the "first to publish, not first to keep secret" rule, the credit of the invention of public-key cryptography goes to Diffie, Hellman and Merkle for their seminar idea and to Rivest, Shamir and Adleman for their first implementation. The claims from CESG/GCHQ and NSA, however, are interesting footnote to the history of modern cryptography.

RSA is now the most popular cryptosystem used to safe-guard our private communications and/or business transactions over the insecure Internet. Readers are suggested to consult the following references for more information about RSA cryptography and cryptnanlysis: Buchmann [60], Delfs and Knebl [97], Wagstaff Jr. [321], Koblitz [172], [174], [175], Koblitz and Menezes [176], Konheim [179], Mollin [216], Mao [196], Menezes et al [207], Muller [225], Rothe [269], Salomann [274], Schneier [277], and Stamp and Low [310], Stinson [311], de Weger [323].

There are many variants of RSA, and there are many other types of cryptosystems whose security are also rely on the intractability of integer factorization problem or the related discrete logarithm problem. Notable systems on these directions include but are not limited to, Rabin's M^2 mod N system [252], Williams M^3 mod N system [327], ElGamal discrete logarithm system [106], Koblitz [170] and Miller [213] elliptic curve discrete logarithm systems, Goldwasser and Micali's quadratic residuosity system [124], and Goldwasser, Micali and Rackoff's zero-knowledge proof system [125].

3. Integer Factorization Attacks

Of all the problems in the theory of numbers to which computers have been applied, probably none has been influenced more than of factoring.

HUGH C. WILLIAMS
Professor of Computer Science, University of Calgary

3.1 Introduction

Recall that the RSA encryption and decryption are defined as follows:

$$\begin{aligned} E_e(M) &= C \equiv M^e \pmod{N}, \\ D_d(C) &= M \equiv C^d \pmod{N} \end{aligned}$$

where $ed \equiv 1 \pmod{\phi(N)}$. Recall also that the RSA problem as follows:

$$\{e, N = pq, C \equiv M^e \pmod{N}\} \xrightarrow{\text{find}} \{M \equiv C^d \pmod{N}\},$$

where p and q are two large distinct primes. Clearly, anyone who can factor N can solve RSAP (and hence break RSA) by computing the inverse of the RSA function $M \mapsto M^e$:

$$M \equiv C^{1/e \pmod{(p-1)(q-1)}} \pmod{N}.$$

That is, if the prime factorization of N is given, then RSAP can be solved in polynomial-time.

$$\text{IFP} \overset{\mathcal{P}}{\Longrightarrow} \text{RSAP}.$$

However, it is not known whether or not

$$\text{IFP} \stackrel{\mathcal{P}}{\Longleftarrow} \text{RSAP}$$

although it is conjectured to be true. Recent research indicates however that breaking RSA may be easier than factoring [40], but there is also evidence to indicate that breaking RSA may be as hard as factoring [57] in general.

Consequently, the most direct method of breaking RSA is to factor the RSA modulus N. Generally speaking, the most useful factoring algorithms fall into one of the following two main categories ([46]):

(I) The running time depends mainly on the size of N, the number to be factored, and is not strongly dependent on the size of the factor p found. Examples are:

(1) *Lehman's method* [185], which has a rigorous worst-case running time bound $\mathcal{O}\left(N^{1/3+\epsilon}\right)$.

(2) *Shanks' SQUare FOrms Factorization method* SQUFOF[199], which has expected running time $\mathcal{O}\left(N^{1/4}\right)$.

(3) *Continued FRACtion (*CFRAC*) method* [223], which under plausible assumptions has expected running time

$$\mathcal{O}\left(\exp\left(c\sqrt{\log N \log\log N}\,\right)\right) = \mathcal{O}\left(N^{c\sqrt{\log\log N/\log N}}\,\right),$$

where c is a constant (depending on the details of the algorithm); usually $c = \sqrt{2} \approx 1.414213562$.

(4) *Quadratic Sieve/Multiple Polynomial Quadratic Sieve (*QS/MPQS*)* [248], which under plausible assumptions has expected running time

$$\mathcal{O}\left(\exp\left(c\sqrt{\log N \log\log N}\,\right)\right) = \mathcal{O}\left(N^{c\sqrt{\log\log N/\log N}}\,\right),$$

where c is a constant (depending on the details of the algorithm), usually $c = \dfrac{3}{2\sqrt{2}} \approx 1.060660172$.

(5) *Number Field Sieve (*NFS*)* [187], which under plausible assumptions has the expected running time

$$\mathcal{O}\left(\exp\left(c\sqrt[3]{\log N}\sqrt[3]{(\log\log N)^2}\,\right)\right),$$

where $c = (64/9)^{1/3} \approx 1.922999427$ if GNFS (a general version of NFS) is used to factor an arbitrary integer N, whereas $c = (32/9)^{1/3} \approx 1.526285657$ if SNFS (a special version of NFS) is used to factor a special integer N such as $N = r^e \pm s$, where r and s are small, $r > 1$ and e is large. This is substantially and asymptotically faster than any other currently known factoring method.

(II) The running time depends mainly on the size of p (the factor found) of N. (We can assume that $p \le \sqrt{N}$.) Examples are:

(1) *Trial division*, which has running time $\mathcal{O}\left(p(\log N)^2\right)$.

(2) *Pollard's "$p-1$" method* [242], which has running time

$$\mathcal{O}\left(B \log B \log N)^2\right),$$

where B is the bound of the Factor Base of small prime numbers.

(3) *Pollard's ρ-method* [243] (also known as Pollard's "*rho*" algorithm), which under plausible assumptions has expected running time $\mathcal{O}\left(p^{1/2}(\log N)^2\right)$.

(4) *Lenstra's Elliptic Curve Method (*ECM*)* [186], which under plausible assumptions has expected running time

$$\mathcal{O}\left(\exp\left(c\sqrt{\log p \log\log p}\,\right)\cdot(\log N)^2\right),$$

where $c \approx 2$ is a constant (depending on the details of the algorithm).

The term $\mathcal{O}\left((\log N)^2\right)$ is a generous allowance for the cost of performing arithmetic operations on numbers which are $\mathcal{O}(\log N)$ or $\mathcal{O}\left((\log N)^2\right)$ bits long; these could theoretically be replaced by $\mathcal{O}\left((\log N)^{1+\epsilon}\right)$ for any $\epsilon > 0$.

In practice, algorithms in both categories are important and useful. However, for attacking RSA with modulus N, only some of the factoring methods are useful.

Of course, there are also some other types of classifications of factoring algorithms. For example, we can classify all factoring algorithms into *deterministic* and *non-deterministic* categories; for instance, trial division factoring algorithm is deterministic whereas the number field sieve is not deterministic. We can also classify all factoring algorithms into *conditional* and *non-conditional* categories. By *conditional* we mean that some pre-condition may be given to the integer N to be factorized. For instance, let $N = pq$ be an odd composite with β bits, p a prime with $\beta/2$ bits and q a prime also with $\beta/2$ bits, then given $\beta/4$ most significant (or least significant) bits of p (or q), then there is an algorithm to factor N in polynomial-time. In this chapter, we shall study various factoring methods that are particularly useful and applicable for breaking the RSA cryptosystem with modulus $N = pq$.

3.2 Fermat Factoring Attack

For N to be hard to factor or for RSA to be difficult to break, p and q should be choose to have the same bit size. However, if the p and q are too close to each other, then Fermat's factoring method could find them. Let $N = pq$, where $p \leq q$ are both odd, then by setting $x = \frac{1}{2}(p+q)$ and $y = \frac{1}{2}(q-p)$ we find that $N = x^2 - y^2 = (x+y)(x-y)$, or $y^2 = x^2 - N$. A version of Fermat's factoring algorithm is given as follows, based on the above idea.

Algorithm 3.2.1 (Fermat's factoring algorithm). Given an odd integer $N > 1$, then this algorithm determines the largest factor $\leq \sqrt{N}$ of N.

[1] Input n and set $k \leftarrow \lfloor \sqrt{N} \rfloor + 1$, $y \leftarrow k \cdot k - N$, $d \leftarrow 1$

[2] If $\lfloor \sqrt{y} \rfloor = \sqrt{y}$ goto step [4] else $y \leftarrow y + 2 \cdot k + d$ and $d \leftarrow d + 2$

[3] If $\lfloor \sqrt{y} \rfloor < N/2$ goto step [2] else print "No Factor Found" and goto [5]

[4] $x \leftarrow \sqrt{N+y}$, $y \leftarrow \sqrt{y}$, print $x - y$ and $x + y$, the nontrivial factors of N

[5] Exit: terminate the algorithm.

Theorem 3.2.1. Let $N = pq$ with $p > q$ and $\Delta = p - q$. Then when $\Delta < N^{1/4}$, Fermat's factoring algorithm can factor N efficiently.

Proof. See de Weger [323]. □

Wagstaff [321] also suggested to choose $|q - p| > 10^{25}$.

3.3 The "$p \pm 1$" and ECM Attacks

The prime numbers p and q in RSA should also to be chosen with the properties that $p \pm 1$ and $q \pm 1$ have at least one prime factor greater than 10^{20} [321], otherwise, p could be found efficiently by using Pollard's "$p - 1$" factoring algorithm [242] and Williams' "$p + 1$" factoring algorithm [329].

Algorithm 3.3.1 ("$p - 1$" factoring). Let $N > 1$ be a composite number. This algorithm attempts to find a nontrivial factor of N.

[1] (Initialization) Choose $a \in \mathbb{Z}_N$ at random. Select a positive integer k that is divisible by many prime powers, for example, $k = \text{lcm}(1, 2, \cdots, B)$ for a suitable bound B (the larger B is the more likely the method will be to succeed in producing a factor, but the longer the method will take to work).

[2] (Exponentiation) Compute $a_k \equiv a^k \pmod{N}$.

[3] (Compute GCD) Compute $f = \gcd(a_k - 1, N)$.

[4] (Factor Found?) If $1 < f < N$, then f is a nontrivial factor of N, output f and go to [6].

[5] (Start Over?) If f is not a nontrivial factor of N and if you still want to try more experiments, then go to [2] to start all over again with a new choice of a and/or a new choice of k, else goto [6].

[6] (Exit) Terminate the algorithm.

The complexity of the algorithm is $\mathcal{O}\left(B \log B (\log N)^2\right)$, so the algorithm is useful only when a small value of B is chosen.

Example 3.3.1. Use the "$p-1$" method to factor the number $N = 540143$. Choose $B = 8$ and hence $k = \text{lcm}(1,2,3,4,5,6,7,8) = 840$. Choose also $a = 2$. Then

$$\gcd(2^{840} - 1 \bmod 540143,\ 540143) = \gcd(53046,\ 540143) = 421.$$

In fact, $540143 = 421 \cdot 1283$. Note that all the prime factors of 421 are small as $421 - 1 = 2^2 \cdot 3 \cdot 5 \cdot 7$, the "$p-1$" method is well suited for splitting 421 from 540143.

Example 3.3.2. The most successful example of using the $p-1$ method is that in 1980 Baillie [48] used it to find the prime factor p_{25} of the Mersenne number

$$2^{257} - 1 = p_{15} \cdot p_{25} \cdot p_{39},$$

where

$$\begin{aligned} p_{15} &= 535006138814359, \\ p_{25} &= 1155685395246619182673033, \text{ and} \\ p_{39} &= 374550598501810936581776630096313181393. \end{aligned}$$

Since

$$p_{25} - 1 = 2^3 \cdot 3^2 \cdot 19^2 \cdot 47 \cdot 67 \cdot 257 \cdot 439 \cdot 119173 \cdot 1050151$$

is a smooth number, the "$p-1$" method is good at finding the p_{25} of $2^{257} - 1$. It is interesting to note that $2^{257} - 1$ was claimed to be prime by Mersenne.

The "$p-1$" algorithm is usually successful in the fortunate case where N has a prime divisor p for which $p-1$ has no large prime factors (i.e, $p-1$ is a smooth number). Suppose that $(p-1) \mid k$ and that $p \nmid a$. Since $|\mathbb{Z}_p^*| = p-1$, we have $a^k \equiv 1 \pmod{p}$, thus $p \mid \gcd(a_k - 1,\ N)$. In many cases, we have $p = \gcd(a_k - 1,\ N)$, so the method finds a nontrivial factor of N. In the worst case, where $(p-1)/2$ is prime, the "$p-1$" algorithm is no better than the trial division method. Since the group has fixed order $p-1$ there is nothing to be done except try a different algorithm.

Note that there is a similar method to "$p-1$", called "$p+1$", proposed by H. C. Williams [329] in 1982. It is suitable for the case where N has a prime factor p for which $p+1$ has no large prime factors. Thus, both "$p-1$" and "$p+1$" will be useful attack on RSA if p or q in the RSA modulus N have the property that $p+1$ or $q+1$ is a smooth number. If $p-1$ is not smooth, Algorithm 3.3.1 will eventually be successful if a large B is chosen but it will not be efficient and practical.

Note that the famous Lenstra's Elliptic Curve factoring Method (ECM) [186] is actually obtained from Pollard's "$p-1$" algorithm: if we can choose a *random* group G with order g close to p, we may be able to perform a computation similar to that involved in Pollard's "$p-1$" algorithm, working in G rather than in F_p. If all prime factors of g are less than the bound B then we find a factor of N. Otherwise, we repeat this procedure with a different group G (and hence, usually, a different g) until a factor is found. This is the whole idea and also the motivation of the ECM method.

Algorithm 3.3.2 (Lenstra's Elliptic Curve Method). Let $N > 1$ be a composite number, with $\gcd(N, 6) = 1$. This algorithm attempts to find a nontrivial factor of N. The method uses elliptic curves and is analogous to Pollard's "$p-1$" method.

[1] (Choose an Elliptic Curve) Choose a random pair (E, P), where E is an elliptic curve $y^2 = x^3 + ax + b$ over $\mathbb{Z}_N$, and $P(x, y) \in E(\mathbb{Z}_N)$ is a point on E. That is, choose $a, x, y \in \mathbb{Z}_N$ at random, and set $b \leftarrow y^2 - x^3 - ax$. If $\gcd(4a^3 + 27b^2, N) \neq 1$, then E is not an elliptic curve, start all over and choose another pair (E, P).

[2] (Choose an Integer k) Just as in the "$p-1$" method, select a positive integer k that is divisible by many prime powers, for example, $k = \text{lcm}(1, 2, \cdots, B)$ or $k = B!$ for a suitable bound B; the larger B is the more likely the method will succeed in producing a factor, but the longer the method will take to work.

[3] (Calculate kP) Calculate the point $kP \in E(\mathbb{Z}/N\mathbb{Z})$. We use the following formula to compute $P_3(x_3, y_3) = P_1(x_1, y_1) + P_2(x_2, y_2) \bmod N$:

$$(x_3, y_3) = (\lambda^2 - x_1 - x_2 \mod N, \quad \lambda(x_1 - x_3) - y_1 \mod N),$$

where

$$\lambda = \begin{cases} \dfrac{m_1}{m_2} \equiv \dfrac{3x_1^2 + a}{2y_1} \pmod N & \text{if } P_1 = P_2 \\[2ex] \dfrac{m_1}{m_2} \equiv \dfrac{y_2 - y_1}{x_2 - x_1} \pmod N & \text{otherwise.} \end{cases}$$

The computation of $kP \bmod N$ can be done in $\mathcal{O}(\log k)$ doublings and additions.

[4] (Compute GCD) If $kP \equiv \mathcal{O}_E \pmod N$, then set $m_2 = z$ and compute $d = \gcd(z, N)$, else goto [1] to make a new choice for "a" or even for a new pair (E, P).

[5] (Factor Found?) If $1 < d < N$, then d is a nontrivial factor of N, output d and go to Step [7].

[6] (Start Over?) If d is not a nontrivial factor of N and if you still wish to try more elliptic curves, then go to Step [1] to start all over again, else go to Step [7].

[7] (Exit) Terminate the algorithm.

As for the "$p-1$" method, one can show that a given pair (E, P) is likely to be successful in the above algorithm if N has a prime factor p for which $\mathbb{Z}_p$ is composed of small primes only. The probability for this successful case happening increases with the number of pairs (E, P) that one tries.

Example 3.3.3. Use the ECM method to factor the number $N = 187$.

[1] Choose $B = 3$, and hence $k = \text{lcm}(1, 2, 3) = 6$. Let $P = (0, 5)$ be a point on the elliptic curve $E: y^2 = x^3 + x + 25$ which satisfies $\gcd(N, 4a^3 + 27b^2) = \gcd(187, 16879) = 1$ (note that here $a = 1$ and $b = 25$).

[2] Since $k = 6 = 110_2$, we compute $6P = 2(P + 2P)$ in the following way:

[2-1] Compute $2P = P + P = (0, 5) + (0, 5)$:

$$\begin{cases} \lambda = \dfrac{m_1}{m_2} = \dfrac{1}{10} \equiv 131 \pmod{187} \\ x_3 = 144 \pmod{187} \\ y_3 = 18 \pmod{187}. \end{cases}$$

So, $2P = (144, 18)$ with $m_2 = 10$ and $\lambda = 131$.

[2-2] Compute $3P = P + 2P = (0, 5) + (144, 18)$:

$$\begin{cases} \lambda = \dfrac{m_1}{m_2} = \dfrac{13}{144} \equiv 178 \pmod{187} \\ x_3 = 124 \pmod{187} \\ y_3 = 176 \pmod{187}. \end{cases}$$

So, $3P = (124, 176)$ with $m_2 = 144$ and $\lambda = 178$.

[2-3] Compute $6P = 2(3P) = 3P + 3P = (124, 176) + (124, 176)$:

$$\lambda = \frac{m_1}{m_2} = \frac{46129}{352} \equiv \frac{127}{165} \equiv \mathcal{O}_E \pmod{187}.$$

This time $m_1 = 127$ and $m_2 = 165$, so the modular inverse for $127/165$ modulo 187 does not exist; but this is exactly what we want! – this type of failure is called a "pretended failure". We now set $z = m_2 = 165$.

[3] Compute $d = \gcd(N, z) = \gcd(187, 165) = 11$. Since $1 < 11 < 187$, 11 is a (prime) factor of 187. In fact, $187 = 11 \cdot 17$.

The factoring methods discussed in this section are only useful either when the RSA modulus N happens to have a prime factor or when $N = pq$, $p \pm 1$ or $q \pm 1$ of N have small prime factors. These special cases (in fact, faults) in the design of RSA can easily be avoided, so the next two sections will be on the powerful general factoring methods such as the Quadratic Sieve and the Number Field Sieve.

3.4 Quadratic Sieve Attack

The Quadratic Sieve (QS), invented by Pomerance (then at the University of Georgia) in 1981 and first published in 1982 [245], belongs to a wide range of factoring algorithms, called index calculus of factoring, along with Continued FRACtion method (CFRAC) [223] and Number Field Sieve (NFS) [187]; all of them make use of the simple but important observation that if we have two integers x and y such that

$$x^2 \equiv y^2 \pmod{N}, \ 0 < x < y < N, \ x \neq y, \ x + y \neq N, \tag{3.1}$$

then $\gcd(x \pm y, \ N)$ are possibly the nontrivial factors of N, because $N \mid (x+y)(x-y)$, but $N \nmid (x+y)$ and $N \nmid (x-y)$. For example, to factor $N = 8051$, we find $90^2 \equiv 7^2 \pmod{N}$, hence $\gcd(90 \pm 7, \ N) = (97, 83)$, thus $8051 = 83 \cdot 97$. How to find the x and y such that the congruence (3.1) is satisfied is the main task of the index calculus; different methods use different techniques to find such pairs of (x, y). A version of QS may be described as follows:

Algorithm 3.4.1 (Quadratic Sieve). Let N be an odd composite that is not a power. This algorithm attempts to find a nontrivial factor f of N such that $f \mid N$ and $1 < f < N$.

[1] (Factor Base) Define a factor base as follows:

$$\text{FB} = \{-1, p_1, p_2, \cdots, p_k \leq B\}$$

where p_i are primes for which N is a quadratic residue modulo p_i, and B is the upper bound of the factor base (the largest prime in the factor base).

[2] (Smoothness) Find $a_1, a_2, \cdots, a_k$, close to $\sqrt{N}$ (this can be done via e.g., $a_i = \lfloor \sqrt{N} \rfloor + 1, \lfloor \sqrt{N} \rfloor + 2, \cdots, (N-1)/2$) such that each $Q(a_i) = a_i^2 - N$ is smooth (a number is smooth if all its prime factors are small with respect to the bound B. In this case, the number is called B-smooth).

[3] (Linear Algebra – Finding $x^2 \equiv y^2 \pmod{N}$) Use linear algebra to find a subset U of the numbers $Q(a_i) = a_i^2 - N$ whose product $\prod p_i^{\alpha_i}$ is a square, say $y^2 \bmod N$. That is, $y^2 \equiv \prod a_i^2 - N$. Let x be the product a_i used to form the square, modulo N. Then

$$
\begin{aligned}
x^2 &\equiv \left(\prod_{i \in U} a_i\right)^2 \\
&\equiv \prod_{i \in U} (a_i^2 - N) \\
&\equiv \prod_{i \in U} Q(a_i) \\
&\equiv \left(\prod_{i \in U} p_j^{\alpha_{j,i}}\right)^2 \\
&\equiv y^2 \qquad \pmod N.
\end{aligned}
$$

[4] (Computing GCD) $(f, g) = \gcd(x \pm y, N)$.

[5] (OK?) If $1 < f, g < N$, print (f, g) (in terms of RSA, (f, g) will be the prime factors (p, q) of the modulus N) and go to [6]. Otherwise, go to [3] to find new x and y and. If necessary, go to [2] to find more a_i's.

[6] Exit.

Example 3.4.1. Use Algorithm 3.4.1 to factor $N = 1829$.

[1] (Factor Base) Let the factor base be as follows:

$$\text{FB} = \{-1, 2, 5, 7, 11\}.$$

Note although $3 < 11$ is a prime but for which N is not a quadratic residue, so we exclude it from the factor base.

[2] (Smoothness) Choose $a_i \sim \lfloor\sqrt{1829}\rfloor = 29$. Let $a_i = 27, 28, 29, \cdots$, compute $Q(a_i) = a_i^2 - N$, keep only the smooth $Q(a_i)$, and get the corresponding exponent vectors modulo 2 as follows:

			-1 2 5 7 11
1)	$Q(27) = 27^2 - N = -1100 = -2^2 \cdot 5^2 \cdot 11$	$\longleftrightarrow$	(1, 0, 0, 0, 1)
2)	$Q(38) = 38^2 - N = -385 = -5 \cdot 7 \cdot 11$	$\longleftrightarrow$	(1, 0, 1, 1, 1)
3)	$Q(39) = 39^2 - N = -308 = -2^2 \cdot 7 \cdot 11$	$\longleftrightarrow$	(1, 0, 0, 1, 1)
4)	$Q(43) = 43^2 - N = 20 = 2^2 \cdot 5$	$\longleftrightarrow$	(0, 1, 1, 0, 0)
5)	$Q(45) = 45^2 - N = 196 = 2^2 \cdot 7^2$	$\longleftrightarrow$	(0, 0, 0, 0, 0)
6)	$Q(52) = 52^2 - N = 875 = 5^3 \cdot 7$	$\longleftrightarrow$	(0, 0,1 , 1,0)
7)	$Q(53) = 53^2 - N = 980 = 2^2 \cdot 5 \cdot 7^2$	$\longleftrightarrow$	(0, 0, 1, 0, 0)

[3] (Linear Algebra – Finding $x^2 \equiv y^2 \pmod N$) Use linear algebra to find a subset of the numbers $Q(a_i) = a_i^2 - n$ whose product $\prod p_i^{\alpha_i}$ is a square; if the sum of the corresponding exponent vectors modulo 2 is zero, then the subset of the numbers $Q(a_i)$ form a square. Observe that (this can be done systematically) the sum of the first, the second and the sixth vectors is zero. That is,

$$\begin{array}{rcl} & (1,0,0,0,1) & \text{1st} \\ & (1,0,1,1,1) & \text{2nd} \\ \oplus & (0,0,1,1,0) & \text{6th} \\ \hline & (0,0,0,0,0,0) & \Longrightarrow \text{Successful} \end{array}$$

So, we have found a suitable pair of (x, y), which produce squares in both sides $(27 \cdot 38 \cdot 52)^2 \equiv (2 \cdot 5^3 \cdot 7 \cdot 11)^2$. Thus, we have

$$\begin{aligned} x &= 27 \cdot 38 \cdot 52 \\ &= 53352 \\ &\equiv 311 \pmod{1829} \\ y &= 2 \cdot 5^3 \cdot 7 \cdot 11 \\ &= 19250 \\ &\equiv 960 \pmod{1829}. \end{aligned}$$

Note that some other subsets of the number $Q(a_i)$ such as the 2nd, 3rd and 7th also form a square:

$$\begin{array}{rcl} & (1,0,1,1,1) & \text{2nd} \\ & (1,0,0,1,1) & \text{3rd} \\ \oplus & (0,0,1,0,0) & \text{7st} \\ \hline & (0,0,0,0,0,0) & \Longrightarrow \text{Successful} \end{array}$$

That is, $(38 \cdot 39 \cdot 53)^2 \equiv (2^2 \cdot 5 \cdot 7^2 \cdot 11)^2$. Thus,

$$\begin{aligned} x &= 38 \cdot 39 \cdot 53 \\ &= 78546 \\ &\equiv 1728 \pmod{1829} \\ y &= 2^2 \cdot 5 \cdot 7^2 \cdot 11 \\ &= 10780 \\ &\equiv 1635 \pmod{1829}. \end{aligned}$$

[4] (Computing GCD) Compute $(f, g) = \gcd(x \pm y, N)$, and hopefully, (f, q) will be the required prime factors (p, q) of N. Since we have found two pairs of

$$(x, y) = (311, 960) = (1728, 1635).$$

Thus we have

$$(f, g) = \gcd(x \pm y, N) = \gcd(311 \pm 960, 1829) = (31, 59).$$

That is, $1829 = 31 \cdot 59$. Alternatively, we have

$$(f, g) = \gcd(x \pm y, N) = \gcd(1728 \pm 1635, 1829) = (59, 31).$$

That is, $1829 = 59 \cdot 31$.

There are many tricks to enhance the performance of the basic QS. Among these are the large and small prime variations and the use of a multiplier. But by far the most important is the multiple polynomial variation, the Multiple Polynomial Quadratic Sieve (MPQS). Different versions of MPQS have been suggested, independently by Davis and Holdridge [94], and Montgomery [246], with Montgomery's slightly better.

In what follows, we shall introduce a version of the MPQS algorithm, due to Riele et al [259]. The idea of the MPQS is as follows. To find the (x, y) pair in

$$x^2 \equiv y^2 \pmod N \tag{3.2}$$

we try to find triples (U_i, V_i, W_i), for $i = 1, 2, \cdots$, such that

$$U_i^2 \equiv V_i^2 W_i \pmod N \tag{3.3}$$

where W is easy to factor (at least easier than N). If suffiently many congruences (3.3) are found, they can be combined, by multiplying together a subset of them, in order to get a relation of the form (3.2). The version of the MPQS algorithm described here is based on [259].

Algorithm 3.4.2 (Multiple Polynomial Quadratic Sieve). Given a positive integer $N > 1$, this algorithm will try to find a factor N using the multiple polynomial quadratic sieve.

[1] Choose B and M, and compute the factor base FB.
Note: M is some fixed integer so that we can define: $U(x) = a^2x + b$, $V = a$ and $W(x) = a^2x^2 + 2bx + c$, $x \in [-M, M)$, such that a, b, c satisfy the following relations:

$$a^2 \approx \sqrt{2N/M}, \quad b^2 - N = a^2c, \quad |b| < (a^2)/2. \tag{3.4}$$

Note: Since the potential prime divisors p of a given quadratic polynomial $W(x)$ may be characterized as: if $p \mid W(x)$, then

$$a^2W(x) = (a^2x + b)^2 - N \equiv 0 \pmod p. \tag{3.5}$$

That is, the congruence $t^2 - N \equiv 0 \pmod p$ should be solvable. So, the factor base FB (consisting of all primes less than a bound B) should be chosen in such a way that $t^2 \equiv N \pmod p$ is solvable. There are L primes p_j, $j = 1, 2, \cdots, L$ in FB; this set of primes is fixed in the whole factoring process.

[2] Generate a new quadratic polynomial $W(x)$.
Note: The quadratic polynomial $W(x)$ *in*

$$(U(x))^2 \equiv V(x)W(x) \pmod N \tag{3.6}$$

assumes extreme values in $x = 0, \pm M$ *such that* $|W(0)| \approx |W(\pm M)| \approx M\sqrt{N/2}$. *If* $M \ll N$, *then* $W(x) \ll N$, *thus* $W(x)$ *is easier to factor than* N.

[3] Solve $W(x) \equiv 0 \pmod q$ for all $q = p^e < B$, for all primes $p \in \mathrm{FB}$, and save the solutions for each q.

[4] Initialize the sieving array $\mathrm{SI}[-M, M)$ to zero.

[5] Add $\log p$ to all elements $\mathrm{SI}(j)$, $j \in [-M, M]$, for which $W(j) \equiv 0 \pmod q$, for all $q = p^e < B$, and for all primes $p \in P\mathrm{FB}$.
Note: Now we can collect those $x \in [-M, M)$ *for which* $W(x)$ *is only composed of prime factors* $< B$.

[6] Selecting those $j \in [-M, M)$ for which $\mathrm{SI}(j)$ is close to $\log(M/2\sqrt{N/2})$.

[7] If the number of $W(x)$-values collected in Step 6 is $< L+2$, then go to Step 2 to construct a new polynomial $W(x)$.
Note: If at least $L+2$ *completely factorized* W*-values have been collected, then the* (x, y)*-pairs satisfying (3.2) may be found as follows. For* x_i, $i = 1, 2, \cdots, L+2$,

$$W(x_i) = (-1)^{\alpha_{i0}} \prod_{j=1}^{L} p_j^{\alpha_{ij}}, \quad i = 1, 2, \cdots, L+2. \tag{3.7}$$

[8] Perform Gaussian elimination on the matrix of exponents (mod 2) of $W(x)$.
Note: Associated with each $W(x_i)$, *we define the vector* α_i *as follows*

$$\alpha_i^T = (\alpha_{i0}, \alpha_{i1}, \cdots, \alpha_{iL}) \pmod 2. \tag{3.8}$$

Since we have more vectors α_i *(at least* $L+2$*) than compoenets* $(L+1)$, *there exists at least one subset* S *of the set* $\{1, 2, \cdots, L+2\}$ *such that*

$$\sum_{i \in S} \alpha_i \equiv 0 \pmod 2,$$

so that

$$\prod_{i \in S} W(x) = Z^2.$$

Hence, from (3.6) it follows that

$$\left[\prod_{i \in S} (a^2 x_i + b)\right] \equiv Z^2 \prod_{i \in S} a^2 \pmod N$$

which is of the required form $x^2 \equiv y^2 \pmod N$.

[9] Factor N.
Note: Now we can calculate $\gcd(x \pm y, N)$ *to find the prime factors of* N.

Conjecture 3.4.1 (The complexity of the QS/MPQS Method). *If N is the integer to be factored, then under certain reasonable heuristic assumptions, the QS/MPQS method will factor N in time*

$$\mathcal{O}\left(\exp\left((1+o(1))\sqrt{\log N \log\log N}\right)\right) = \mathcal{O}\left(N^{(1+o(1))\sqrt{\log\log N/\log N}}\right).$$

The MPQS is however not the fastest factoring algorithm at present; the fastest factoring algorithm in use today is the Number Field Sieve, which is the subject matter of our next section.

3.5 Successful QS Attack

The most successful QS attack on RSA is the following challenge, proposed by RSA but first appeared in Martin Gardner's 1977 paper [115]:

9686	9613	7546	2206
1477	1409	2225	4355
8829	0575	9991	1245
7431	9874	6951	2093
0816	2982	2514	5708
3569	3147	6622	8839
8962	8013	3919	9055
1829	9451	5781	5154

A Ciphertext Challenge Worth $100

The text "THE MAGIC WORDS ARE SQUEAMISH OSSIFRAGE" was the solution to the challenge. The idea behind this challenge is that it needs to factor a 129-digit number, RSA-129 first (see Example 2.3.1). Rivest estimated in 1977 [115] that factoring a 125-digit number would require 40 quadrillion years using the best algorithm and fastest computer at that time; he therefore then believed that RSA-129 could never be factored and hence the above challenge could never be broken. What Rivest failed to take into account was the possibility of progress in integer factorization. This section gives a brief description of the successful attack on this challenge using QS.

Notice first that for

$$M \equiv C^d \pmod N, \quad d \equiv 1/e \pmod{\phi(N)}$$

C is the following number from the proposed challenge problem:

96869613754622061477140922254355882905759991124574319874695120930_
08162982251457083569314766228839896280133919905518299451578151 54

(e, N) are public with $e = 9007$ and N the following number:

114381625757888867669235779976146612010218296721242362562561842 9_
35706935245733897830597123563958705058989075147599290026879543541.

So, it is trivial to find M if the prime factorization of $N = pq$ can be found since $\phi(N) = (p-1)(q-1)$. In 1994, 13 years later, N was factored using MPQS; more than 600 volunteers across America, Europe, Asia, and Australia, contributed CPU time from about 1600 machines over 6 months. The two prime factors are as follows:

3490529510847650949147849619903898133417764638493387843990820577,
32769132993266709549961988190834461413177642967992942539798288533.

Now, $d = 1/e \pmod{(p-1)(q-1)}$ can be easily calculated as follows:

106698614368578024442868771328920154780709906633937862801226224496631063125911774470873340168597462306553968544513277109053606095.

Therefore, $M \equiv C^d \pmod{N}$ can be easily calculated via

$$
\begin{array}{l}
M \leftarrow 1 \\
\text{while } d \geq 1 \text{ do} \\
\qquad \text{if } d \bmod 2 = 1 \\
\qquad \quad \text{then } M \leftarrow C \cdot M \bmod N \\
\qquad C \leftarrow C^2 \bmod N \\
\qquad d \leftarrow \lfloor d/2 \rfloor \\
\text{print } M \text{ (Now } M \equiv C^d \pmod{N})
\end{array}
$$

which gives the plaintext M:

200805001301070903002315180419000118050019172105011309190800151919090618010705.

In what follows, we show how to factor N using MPQS [15].

[1] (Sieving Step) Find a multiplier m such that mN is a quadratic residue modulo many small primes. Then compute the factor base FB, consisting -1 and all primes $p \leq B$ for which mN is a quadratic residue modulo p, for some bound B. The main task in this step is to collect a set of relations (integer tuples $(v, q_1, q_2, (e_p)_{p \in \mathrm{FB}})$ such that

$$v^2 \equiv q_1 \cdot q_2 \cdot \prod_{p \in \mathrm{FB}} p^{e_p} \pmod{N} \tag{3.9}$$

where each q_i is either 1 or a large prime in $(B_1, B_2]$, for some large prime bound B_2. If $q_1 = q_2 = 1$ the relation is called full, if only of q_i equals 1, it is a partial, and a double partial otherwise. If the large primes match, the double partial relations can be combined into cycles. The sieving step is complete as soon as $\#\text{full} + \#\text{cycles} > \#\text{FB}$. For RSA-129, $m = 5$, $B_1 = 16333609$, $\#\text{FB} = 51 + 2^{19}$. It finally had 112011 fulls, 1431337 partial, 6881138 double and 457455 cycles, and led to 569466 aparse 542339-dimensional vectors $(e_p)_{p \in \text{FB}}$.

[2] (Matrix Step) This step is to find linear dependencies modulo 2 among the $>$ FB bit-vectors $(e_p \bmod 2)_{p \in \text{FB}}$. Each dependency corresponds to a set $W = \{(w, (e_p)_{p \in \text{FB}}) : w^2 \equiv \prod_{p \in \text{FB}} p^{e_p} \bmod N\}$ for which $\sum_W (e_p)_{p \in \text{FB}} = (2w_p)_{p \in \text{FB}}$ for integers w_p. Consequently, $x \equiv \prod_W w \bmod N$ and $y \equiv \prod_{p \in \text{FB}} p^{w_p} \bmod N$ satisfy $x^2 \equiv y^2 \bmod N$. It is interesting to note that the first three dependencies led to $N = 1 \cdot N$. On 2 April 1994, the fourth one led to the factorization $N = pq$, with p the 64-digit prime and q the 65-digit prime, as given above.

MPQS is a factoring method very well suited for integers about 130 digits. For integers larger than 130 digits, a big gun will be needed, which is the subject matter of the next section.

3.6 Number Field Sieve Attack

First let us present a useful lemma:

Lemma 3.6.1. Let $f(x)$ is an irreducible monic polynomial of degree d over integers and m an integer such that $f(m) \equiv 0 \pmod N$. Let α be a complex root of $f(x)$ and $\mathbb{Z}[\alpha]$ the set of all polynomials in α with integer coefficients. Then there exists a unique mapping $\Phi : \mathbb{Z}[\alpha] \mapsto \mathbb{Z}_n$ satisfying

(1) $\Phi(ab) = \Phi(a)\Phi(b), \quad \forall a, b \in \mathbb{Z}[\alpha]$;

(2) $\Phi(a + b) = \Phi(a) + \Phi(b), \quad \forall a, b \in \mathbb{Z}[\alpha]$;

(3) $\Phi(za) = z\Phi(a), \quad \forall a \in \mathbb{Z}[\alpha], z \in \mathbb{Z}$;

(4) $\Phi(1) = 1$;

(5) $\Phi(\alpha) = m \pmod N$.

Note first that there are two main types of NFS: GNFS (general NFS) and SNFS (special NFS). As in RSA, the modulus N is chosen to be a general number, not a number with a special form, we only consider GNFS here; whenever NFS is mentioned, it it to mean GNFS, not SNFS. The basic idea of Number Field Sieve (NFS) is then as follows.

[1] Find a monic irreducible polynomial $f(x)$ of degree d in $\mathbb{Z}[x]$, and an integer m such that $f(m) \equiv 0 \pmod N$.

[2] Let $\alpha \in \mathbb{C}$ be an algebraic number that is the root of $f(x)$, and denote the set of polynomials in α with integer coefficients as $\mathbb{Z}[\alpha]$.

[3] Define the mapping (ring homomorphism): $\Phi : \mathbb{Z}[\alpha] \mapsto \mathbb{Z}_n$ via $\Phi(\alpha) = m$ which ensures that for any $f(x) \in \mathbb{Z}[x]$, we have $\Phi(f(\alpha)) \equiv f(m) \pmod{N}$.

[4] Find a finite set U of coprime integers (a, b) such that

$$\prod_{(a,b)\in U} (a - b\alpha) = \beta^2, \quad \prod_{(a,b)\in U} (a - bm) = y^2$$

for $\beta \in \mathbb{Z}[\alpha]$ and $y \in \mathbb{Z}$. Let $x = \Phi(\beta)$. Then

$$\begin{aligned} x^2 &\equiv \Phi(\beta)\Phi(\beta) \\ &\equiv \Phi(\beta^2) \\ &\equiv \Phi\left(\prod_{(a,b)\in U} (a - b\alpha)\right) \\ &\equiv \prod_{(a,b)\in U} \Phi(a - b\alpha) \\ &\equiv \prod_{(a,b)\in U} (a - bm) \\ &\equiv y^2 \qquad \pmod{N} \end{aligned}$$

which is of the required form of the factoring congruence, and hopefully, a factor of N can be found by calculating $\gcd(x \pm y, N)$.

There are many ways to implement the above idea, all of which follow the same pattern as CFRAC and QS/MPQS: by a sieving process one first tries to find congruences modulo N by working over a factor base, and then do a Gaussian elimination over $\mathbb{Z}/2\mathbb{Z}$ to obtain a congruence of squares $x^2 \equiv y^2 \pmod{N}$. We give in the following a brief description of the NFS algorithm [221].

Algorithm 3.6.1. Given an odd positive integer N, the NFS algorithm has the following four main steps in factoring N:

[1] (Polynomials Selection) Select two irreducible polynomials $f(x)$ and $g(x)$ with small integer coefficients for which there exists an integer m such that

$$f(m) \equiv g(m) \equiv 0 \pmod{N} \tag{3.10}$$

The polynomials should not have a common factor over $\mathbb{Q}$.

[2] (Sieving) Let α be a complex root of f and β a complex root of g. Find pairs (a, b) with $\gcd(a, b) = 1$ such that the integral norms of $a - b\alpha$ and $a - b\beta$:

$$N(a - b\alpha) = b^{\deg(f)} f(a/b), \qquad N(a - b\beta) = b^{\deg(g)} g(a/b) \tag{3.11}$$

are smooth with respect to a chosen factor base. (The principal ideals $a - b\alpha$ and $a - b\beta$ factor into products of prime ideals in the number field $\mathbb{Q}(\alpha)$ and $\mathbb{Q}(\beta)$, respectively.)

[3] (Linear Algebra) Use techniques of linear algebra to find a set S of indices such that the two products

$$\prod_{i \in S} (a_i - b_i \alpha), \qquad \prod_{i \in S} (a_i - b_i \beta) \tag{3.12}$$

are both squares of products of prime ideals.

[4] (Square root) Use the set S in (3.12) to find an algebraic numbers $\alpha' \in \mathbb{Q}(\alpha)$ and $\beta' \in \mathbb{Q}(\beta)$ such that

$$(\alpha')^2 = \prod_{i \in S} (a_i - b_i \alpha), \qquad (\beta')^2 = \prod_{i \in S} (a_i - b_i \beta) \tag{3.13}$$

Define $\Phi_\alpha : \mathbb{Q}(\alpha) \to \mathbb{Z}_N$ and $\Phi_\beta : \mathbb{Q}(\beta) \to \mathbb{Z}_N$ via $\Phi_\alpha(\alpha) = \Phi_\beta(\beta) = m$, where m is the common root of both f and g. Then

$$\begin{aligned}
x^2 &\equiv \Phi_\alpha(\alpha')\Phi_\alpha(\alpha') \\
&\equiv \Phi_\alpha((\alpha')^2) \\
&\equiv \Phi_\alpha\left(\prod_{i \in U} (a_i - b_i \alpha)\right) \\
&\equiv \prod_{i \in U} \Phi_\alpha(a_i - b_i \alpha) \\
&\equiv \prod_{i \in U} (a_i - b_i m) \\
&\equiv \Phi_\beta(\beta')^2 \\
&\equiv y^2 \qquad (\text{mod } N)
\end{aligned}$$

which is of the required form of the factoring congruence, and hopefully, a factor of N can be found by calculating $\gcd(x \pm y, N)$.

Example 3.6.1. Use NFS to factor $N = 1098413$. First notice that $N = 1098413 = 12 \cdot 45^3 + 17^3$, which is in a special form and can be factored by using SNFS.

[1] (Polynomials Selection) Select the two irreducible polynomials $f(x)$ and $g(x)$ and the integer m as follows:

$$m = \frac{17}{45}$$

$$f(x) = x^3 + 12 \implies f(m) = \left(\frac{17}{45}\right)^3 + 12 \equiv 0 \pmod{N}$$

$$g(x) = 45x - 17 \implies g(m) = 45\left(\frac{17}{45}\right)^3 - 17 \equiv 0 \pmod{N}$$

[2] (Sieving)

$$U = \{(6,1), (-3,2), (7,3), (-1,3), (2,5), (3,8), (-9,10)\}.$$

Let $\alpha = \sqrt[3]{-12}$ and $\beta = \frac{17}{45}$. Then

$$\begin{aligned}\prod_U (a - b\alpha) &= 7400772 + 1138236\alpha - 10549\alpha^2 \\ &= (2694 + 213\alpha - 38\alpha^2)^2 \\ &= \gamma^2 \\ \prod_U (a - b\beta) &= \frac{2^8 \cdot 11^2 \cdot 13^2 \cdot 23^2}{3^{12} \cdot 5^4} \\ &= \frac{52624}{18225}\end{aligned}$$

Since

$$\Phi_\alpha(\gamma) \equiv \frac{5610203}{2025} \pmod{N},$$

then

$$\left(\frac{52624}{18225}\right)^2 \equiv \left(\frac{5610203}{2025}\right)^2 \pmod{N}.$$

Thus,

$$\gcd(52624 \cdot 2025 + 5610203 \cdot 18225, 1098413) = (563, 1951).$$

That is,

$$1098413 = 563 \cdot 1951.$$

Conjecture 3.6.1 (Complexity of NFS). Under some reasonable heuristic assumptions, the NFS method can factor an integer N in time

$$\mathcal{O}\left(\exp\left((c + o(1)) \sqrt[3]{\log N} \sqrt[3]{(\log \log N)^2}\right)\right) \tag{3.14}$$

where $c = (64/9)^{1/3} \approx 1.922999427$ if a general version of NFS, GNFS, is used to factor an arbitrary integer N, whereas $c = (32/9)^{1/3} \approx 1.526285657$ if a special version of NFS, SNFS, is used to factor a special form of integer N.

Although the IFP problem is still far way from being solved, significant progress has been made over the last 30 years or so, which can been seen from Table 3.1 of the factorization records (starting from RSA-130, all numbers are factored by Number Field Sieve). In Table 3.1, RSA-xxx

Year	Digits	Bits	RSA-x (digits)	RSA-y (bits)	Prize
1964	20				
1974	45				
1984	71				
1991	100	332	RSA-100		
1992	110	365	RSA-110		
1993	120	398	RSA-120		
1994	129	428	RSA-129		\$100
1996	130	431	RSA-130		
1999	140	465	RSA-140		
1999	155	512	RSA-155		
2003	160	530	RSA-160		
2003	174	576		RSA-576	\$10,000
2005 (Nov)	200	663		RSA-663	
2005 (May)	193	640		RSA-640	\$20,000
?	212	704		RSA-704	\$30,000
?	232	768		RSA-768	\$50,000
?	270	896		RSA-896	\$75,000
?	309	1024		RSA-1024	\$100,000
?	463	1536		RSA-1536	\$150,000
?	617	2048		RSA-2048	\$200,000

Table 3.1. Large Number Factorization Records

(or RSA-yyy) represent the number of digits (or the number of bits) of the RSA number. For example, the number RSA-212 has 212 digits (704 bits):
74037563479561712828046796097429573142593188889231289084936232638972765034028266276891996419625117843995894330502127585370118968098286733173273108930900552505116877063299072396380786710086096962537934650563796359,
whereas RSA-1536 has 1536 bits (463 digits):
184769970321174147430683562020016440301854933866341017147178577491065169671116124985933768430543574458561606154457179405222971773252466096064694607124962372044202226975675668737842756238950876467844093328515749657884341508847552829818672645133986336493190808467199043187438128336350279547028265329780293491615581188104984490831954

500984839377522725705257859194499387007369575568843693381277961308_
923039256969525326162082367649031603655137144791393234716956698806_
9.

3.7 Chapter Notes and Further Reading

Various integer factoring attacks for RSA are discussed in this chapter. Pollard's $p-1$ method [242] and Williams' $p+1$ method [329] are suitable for attacks on RSA modulus N with $p \pm 1$ has no large prime factors. Thus for RSA to be resistant to the $p \pm 1$ attacks, the prime factor p of N must satisfy that $p \pm 1$ have at least one large prime factor. Pollard's ρ attack [243] and H. Lenstra's ECM attack [186] are only suited for attacks on RSA modulus N with a small prime factor. Thus for RSA to be resistant to the ρ attack and ECM attack, the two prime factors p and q of N should be about the same size and cannot be one large and one small. The current most powerful and general methods of factoring are the Quadratic Sieve [248] and Number Field Sieve [187], with QS suitable for factoring a number less than 130 digits and NFS for numbers bigger than 130 digits. The current factoring record of NFS is the RSA-640, a number with 640 bits and 193 digits; it was factored in May 2005, winning the $20,000 prize from RSA Data Security Inc. So, by today's computing technology, for RSA to be resistant to factoring attack, its modulus N must be bigger than 200 digits. In 1990, Pomerance pointed out in [247] that *if one is factoring N so as to cryptanalyze the RSA cryptosystem with modulus N, one might immediately go to QS, skipping ECM completely.* This viewpoint of 1990 must be changed now to *if one is factoring N so as to cryptanalyze the RSA cryptosystem with modulus $N > 10^{130}$, one might immediately go to NFS, skipping QS/MPQS completely.* Of course, if one could build up a practical quantum computer, one should directly go the Shor's quantum factoring algorithm [287], skipping all the classical factoring algorithms, including NFS, completely.

4. Discrete Logarithm Attacks

Many cryptosystems could be broken if we can compute discrete logarithms quickly, that is, if we could solve the equation $a^x = b$ in a large finite field.

SAMUEL S. WAGSTAFF, JR.
Professor of Computer Science at Purdue University

4.1 Introduction

The problem of computing discrete logarithms is fundamental in computational number theory, computational algebra, and of great importance in public-key cryptography

First of all, let us recall the Discrete Logarithm Problem (DLP):

$$\text{DLP}: \quad \{n \in \mathbb{Z}^+_{>1}, x, y, k \in \mathbb{Z}^+, y \equiv x^k \pmod{n}\} \xrightarrow{\text{find}} \{k\}.$$

As we already know DLP is computationally intractable and the security of several well known cryptosystems, such as the Diffie-Hellman-Merkle key-exchange scheme, the ElGammal systems and the US Government's digital signature standard/algorithm (DSS/DSA), is based on the intractability of DLP. So, as for the factoring attack for RSA, the discrete logarithm attack is also the most direct attack for all the DLP-based cryptographic systems. It is interesting to note that the discrete logarithm attack is also the most direct attack for the IFP-based cryptographic systems in general and the RSA cryptographic system in particular. On the one hand, if the adversary Eve has an efficient integer factorization algorithm, she can factor $N = pq$ in polynomial-time, and can compute $d \equiv 1/e \bmod ((p-1)(q-1))$ and hence recover $C \equiv M^d \pmod{N}$. On the other hand, if Eve has an efficient discrete logarithm algorithm, she can make her own plaintext M and use the publically known information (e, N) to get her own ciphertext C for M, then she can use her discrete logarithm algorithm to compute $d \equiv \log_{M^e} M \pmod{N}$, that is,

$$\{N, M, M^e\} \xrightarrow{\text{find}} \{d\}.$$

This is the same as to say that if one can solve the DLP problem in polynomial-time, one can break the RSA system in polynomial-time:

$$\text{DLP} \stackrel{\mathcal{P}}{\Longrightarrow} \text{RSA}.$$

Example 4.1.1. Let the public-key and the cipher-text be as follows (see the RSA original paper [262]):

$$(e, N, C) = (17, 2773, 2258)$$

Find the corresponding M such that $C \equiv M^e \pmod{N}$. The most obvious way to invert the encryption function $M \mapsto C$ is to find d by computing $d \equiv 1/e \pmod{\phi(N)}$ so that $M \equiv C^d \pmod{N}$, but to do so we need to factor N. Now we claim that we can find d in a different way as follows:

[1] Choose a piece of any reasonable plain-text M with $\gcd(N, M) = 1$, say $M = 2113$ and do the following calculation:

$$C \equiv M^e \equiv 2113^{17} \equiv 340 \pmod{2773}.$$

[2] Compute d by taking the discrete logarithm M to the base C modulo N:

$$d \equiv \log_C M \equiv \log_{340} 2113 \equiv 157 \pmod{2773}.$$

[3) Now we can decrypt the given cipher-text $C = 2258$ by using this d:

$$M \equiv C^d \equiv 2258^{157} \equiv 1225 \pmod{2773}.$$

[4] The result can be easily verified to be true, since

$$C \equiv M^e \equiv 1225^{17} \equiv 2258 \pmod{2773}.$$

Example 4.1.2. Let the public-key and the cipher-text be as follows

$$(e, N, C) = (7, 69056069, 19407420)$$

(see page 173 of [218]), find M such that $C \equiv M^e \mod N$). We perform the following computations:

[1] Choose a piece of any reasonable plain-text M with $\gcd(N, M) = 1$, say $M = 59135721$ and do the following calculation:

$$C \equiv M^e \equiv 59135721^7 \equiv 60711351 \pmod{69056069}.$$

[2] Compute d by taking the discrete logarithm M to the based C modulo N:

$$d \equiv \log_C M \equiv \log_{60711351} 59135721 \equiv 4931383 \pmod{69056069}.$$

[3] Also compute e by taking the discrete logarithm C to the based M modulo N:

$$e \equiv \log_C M \equiv \log_{59135721} 60711351 \equiv 7 \pmod{69056069}.$$

[4] Decrypt the given cipher-text $C = 19407420$ by using the d:

$$M \equiv C^d \equiv 19407420^{4931383} \equiv 7289258 \pmod{69056069}.$$

[5] Verify the result using e:

$$M \equiv C^e \equiv 7289258^7 \equiv 19407420 \pmod{69056069}.$$

Now everything this is done, and everything is correct.

Remark 4.1.1. It is very interesting to note that to find the C, one usually needs to compute

$$d \equiv 1/e \pmod{\phi(N)}$$

in order to calculate $C \equiv M^d \pmod{N}$. What we have shown in Example 4.1.2 is that we can find d in an easy way using

$$d \equiv \log_{C'} M' \bmod N$$

in which both M' and C' can be easily calculated. More importantly

[1] The d from M' and C' can be used to decode the C from M.

[2] The d from $\log_{C'} M' \bmod N$ may be different from the d from $1/e \pmod{\phi(N)}$, but they perform exactly the same job in decoding C as can be seen as follows:

$$M \equiv 19407420^{4931383} \equiv 7289258 \pmod{69056069},$$

$$M \equiv 19407420^{39451063} \equiv 7289258 \pmod{69056069}.$$

[3] Our d from $\log_{C'} M' \bmod N$ is smaller than the d from $1/e \pmod{\phi(N)}$ (we might call the d from $1/e \pmod{\phi(N)}$ normal d); this is the sufficient information for the famous RSA system may not be as secure as it claimed; since, et least, we can try many possible d's smaller than the normal d.

Example 4.1.3. Let the public-key and the cipher-text be as follows

$$(e, N, C) = (3533, 11413, 5761)$$

(see page 168 of [311]), find M such that $C \equiv M^e \mod N$). From the given public-key $(e, N) = (3533, 11413)$, it is sufficient to get $d \equiv 1/e \equiv 6597 \pmod{\phi(11413)}$, but we do not want to do so since we assume $\phi(N)$ is hard. We rather do the following computations:

[1] Choose a piece of any reasonable plain-text M' with $\gcd(N, M') = 1$, say $M' = 8611$.

[2] Compute C' as follows:

$$C' \equiv M'^e \equiv 8611^{3533} \equiv 3866 \pmod{11413}.$$

[3] Compute d' as follows:

$$d' \equiv \log_{C'} M' \equiv \log_{3866} 8611 \equiv 997 \pmod{11413}.$$

If we wish, we can find the corresponding e' by

$$e' \equiv \log'_M C \equiv \log_{8611} 3866 \equiv 733 \pmod{11413}.$$

[4] Decrypt the given cipher-text $C = 5761$ by using the d':

$$M \equiv C^{d'} \equiv 5761^{997} \equiv 9726 \pmod{11413}.$$

$$C \equiv M^{e'} \equiv 9726^{733} \equiv 5761 \pmod{11413}.$$

According to Adleman in 1979, the Russian mathematician Bouniakowsky developed a clever algorithm to solve the congruence $a^x \equiv b \pmod{n}$, with the asymptotic complexity $\mathcal{O}(n)$ in 1870. Despite its long history, no efficient algorithm has ever emerged for the discrete logarithm problem. It is believed to be extremely hard, and harder than the integer factorization problem (IFP) even in the average case. The best known algorithm for DLP at present, using NFS and due to Gordon, requires an expected running time

$$\mathcal{O}\left(\exp\left(c(\log n)^{1/3}(\log\log n)^{2/3}\right)\right).$$

There are essentially three different categories of algorithms in use for computing discrete logarithms:

(1) Algorithms that work for arbitrary groups, that is, those that do not exploit any specific properties of groups; Shanks' baby-step giant-step method, Pollard's ρ-method (an analogue of Pollard's ρ-factoring method) and the λ-method (also known as wild and tame Kangaroos) are in this category.

(2) Algorithms that work well in finite groups for which the order of the groups has no large prime factors; more specifically, algorithms that work for groups with smooth orders. A positive integer is called *smooth* if it has no large prime factors; it is called y-smooth if it has no large prime factors exceeding y. The well-known Silver–Pohlig–Hellman algorithm based on the Chinese Remainder Theorem is in this category.

(3) Algorithms that exploit methods for representing group elements as products of elements from a relatively small set (also making use of the Chinese Remainder Theorem); the typical algorithms in this category are Adleman's index calculus algorithm and Gordon's NFS algorithm.

In following sections, we shall introduce the basic ideas of each of these three categories; more specifically, we shall introduce Shanks' baby-step giant-step algorithm, the Silver–Pohlig–Hellman algorithm, Adleman's index calculus algorithm as well as Gordon's NFS algorithm for computing discrete logarithms.

4.2 Baby-Step Giant-Step Attack

Let G be a finite cyclic group of order n, a a generator of G and $b \in G$. The *obvious* algorithm for computing successive powers of a until b is found to take $\mathcal{O}(n)$ group operations. For example, to compute $x = \log_2 15 \pmod{19}$, we compute $2^x \bmod 19$ for $x = 0, 1, 2, \cdots, 19-1$ until $2^x \bmod 19 = 15$ for some x is found, that is:

x	0	1	2	3	4	5	6	7	8	9	10	11
a^x	1	2	4	8	16	13	7	14	9	18	17	15

So $\log_2 15 \pmod{19} = 11$. It is clear that when n is large, the algorithm is inefficient. In this section, we introduce a type of square root algorithm, called the baby-step giant-step algorithm, for taking discrete logarithms, which is better than the above mentioned *obvious* algorithm. The algorithm, due to Daniel Shanks (1917–1996), works on arbitrary groups.

Let $m = \lfloor \sqrt{n} \rfloor$. The baby-step giant-step algorithm is based on the observation that if $x = \log_a b$, then we can uniquely write $x = i + jm$, where $0 \le i, j < m$. For example, if $11 = \log_2 15 \bmod 19$, then $a = 2$, $b = 15$, $m = 5$, so we can write $11 = i + 5j$ for $0 \le i, j < m$. Clearly here $i = 1$ and $j = 2$ so we have $11 = 1 + 5 \cdot 2$. Similarly, for $14 = \log_2 6 \bmod 19$ we can write $14 = 4 + 5 \cdot 2$, for $17 = \log_2 10 \bmod 19$ we can write $17 = 2 + 5 \cdot 3$, etc. The following is a description of the algorithm:

Algorithm 4.2.1 (Shanks' baby-step giant-step algorithm). This algorithm computes the discrete logarithm x of y to the base a, modulo n, such that $y = a^x \pmod{n}$:

[1] (Initialization) Computes $s = \lfloor \sqrt{n} \rfloor$.

[2] (Computing the baby step) Compute the first sequence (list), denoted by S, of pairs (ya^r, r), $r = 0, 1, 2, 3, \cdots, s-1$:

$$S = \{(y, 0), (ya, 1), (ya^2, 2), (ya^3, 3), \cdots, (ya^{s-1}, s-1) \bmod n\} \tag{4.1}$$

and sort S by ya^r, the first element of the pairs in S.

[3] (Computing the giant step) Compute the second sequence (list), denoted by T, of pairs (a^{ts}, ts), $t = 1, 2, 3, \cdots, s$:

$$T = \{(a^s, 1), (a^{2s}, 2), (a^{3s}, 3), \cdots, (a^{s^2}, s) \bmod n\} \tag{4.2}$$

and sort T by a^{ts}, the first element of the pairs in T.

[4] (Searching, comparing and computing) Search both lists S and T for a match $ya^r = a^{ts}$ with ya^r in S and a^{ts} in T, then compute $x = ts - r$. This x is the required value of $\log_a y \pmod{n}$.

This algorithm requires a table with $\mathcal{O}(m)$ entries ($m = \lfloor \sqrt{n} \rfloor$, where n is the modulus). Using a sorting algorithm, we can sort both the lists S and T in $\mathcal{O}(m \log m)$ operations. Thus this gives an algorithm for computing discrete logarithms that uses $\mathcal{O}(\sqrt{n} \log n)$ time and space for $\mathcal{O}(\sqrt{n})$ group elements. Note that Shanks' idea is originally for computing the order of a group element g in the group G, but here we use his idea to compute discrete logarithms. Note also that although this algorithm works on arbitrary groups, if the order of a group is larger than 10^{40}, it will be infeasible.

Example 4.2.1. Suppose we wish to compute the discrete logarithm

$$x = \log_2 6 \bmod 19$$

such that $6 = 2^x \bmod 19$. According to Algorithm 4.2.1, we perform the following computations:

[1] $y = 6$, $a = 2$ and $n = 19$, $s = \lfloor \sqrt{19} \rfloor = 4$.

[2] Computing the baby step:

$$\begin{aligned} S &= \{(y, 0), (ya, 1), (ya^2, 2), (ya^3, 3) \bmod 19\} \\ &= \{(6, 0), (6 \cdot 2, 1), (6 \cdot 2^2, 2), (6 \cdot 2^3, 3) \bmod 19\} \\ &= \{(6, 0), (12, 1), (5, 2), (10, 3)\} \\ &= \{(5, 2), (6, 0), (10, 3), (12, 1)\}. \end{aligned}$$

[3] Computing the giant step:

$$\begin{aligned} T &= \{(a^s, s), (a^{2s}, 2s), (a^{3s}, 3s), (a^{4s}, 4s) \bmod 19\} \\ &= \{(2^4, 4), (2^8, 8), (2^{12}, 12), (2^{16}, 16) \bmod 19\} \\ &= \{(16, 4), (9, 8), (11, 12), (5, 16)\} \\ &= \{(5, 16), (9, 8), (11, 12), (16, 4)\} \end{aligned}$$

[4] Matching and computing: The number 5 is the common value of the first element in pairs of both lists S and T with $r = 2$ and $st = 16$, so $x = st - r = 16 - 2 = 14$. That is, $\log_2 6 \pmod{19} = 14$, or equivalently, $2^{14} \pmod{19} = 6$.

Example 4.2.2. Suppose we wish to find the discrete logarithm

$$x = \log_{59} 67 \bmod 113$$

such that $67 = 59^x \bmod 113$. Again by Algorithm 4.2.1, we have:

[1] $y = 67$, $a = 59$ and $n = 113$, $s = \lfloor \sqrt{113} \rfloor = 10$.

[2] Computing the baby step:

$$\begin{aligned} S &= \{(y, 0), (ya, 1), (ya^2, 2), (ya^3, 3), \cdots, (ya^9, 9) \bmod 113\} \\ &= \{(67, 0), (67 \cdot 59, 1), (67 \cdot 59^2, 2), (67 \cdot 59^3, 3), (67 \cdot 59^4, 4), \\ &\quad (67 \cdot 59^5, 5), (67 \cdot 59^6, 6), (67 \cdot 59^7, 7), (67 \cdot 59^8, 8), \\ &\quad (67 \cdot 59^9, 9) \bmod 113\} \\ &= \{(67, 0), (111, 1), (108, 2), (44, 3), (110, 4), (49, 5), (66, 6), \\ &\quad (52, 7), (17, 8), (99, 9)\} \\ &= \{(17, 8), (44, 3), (49, 5), (52, 7), (66, 6), (67, 0), (99, 9), \\ &\quad (108, 2), (110, 4), (111, 1)\} \end{aligned}$$

[3] Computing the giant-step:

$$\begin{aligned} T &= \{(a^s, s), (a^{2s}, ss), (a^{3s}, 3s), \cdots (a^{10s}, 10s) \bmod 113\} \\ &= \{(59^{10}, 10), (59^{2 \cdot 10}, 2 \cdot 10), (59^{3 \cdot 10}, 3 \cdot 10), (59^{4 \cdot 10}, 4 \cdot 10), \\ &\quad (59^{5 \cdot 10}, 5 \cdot 10), (59^{6 \cdot 10}, 6 \cdot 10), (59^{7 \cdot 10}, 7 \cdot 10), (59^{8 \cdot 10}, 8 \cdot 10), \\ &\quad (59^{9 \cdot 10}, 9 \cdot 10) \bmod 113\} \\ &= \{(72, 10), (99, 20), (9, 30), (83, 40), (100, 50), (81, 60), \\ &\quad (69, 70), (109, 80), (51, 90), (56, 100)\} \\ &= \{(9, 30), (51, 90), (56, 100), (69, 70), (72, 10), (81, 60), (83, 40), \\ &\quad (99, 20), (100, 50), (109, 80)\} \end{aligned}$$

[4] Matching and computing: The number 99 is the common value of the first element in pairs of both lists S and T with $r = 9$ and $st = 20$, so $x = st - r = 20 - 9 = 11$. That is, $\log_{59} 67 \pmod{113} = 11$, or equivalently, $59^{11} \pmod{113} = 67$.

Shanks' baby-step giant-step algorithm is a type of *square root method* for computing discrete logarithms. In 1978 Pollard also gave two other types of square root methods, namely the ρ-method and the λ-method for taking discrete logarithms. Pollard's methods are probabilistic but remove the necessity of precomputing the lists S and T, as with Shanks' baby-step giant-step method. Again, Pollard's algorithm requires $\mathcal{O}(n)$ group operations and hence is infeasible if the order of the group G is larger than 10^{40}.

4.3 Silver–Pohlig–Hellman Attack

In 1978, Pohlig and Hellman proposed an important special algorithm, now widely known as the Silver–Pohlig–Hellman algorithm for computing discrete logarithms over GF(q) with $\mathcal{O}(\sqrt{p})$ operations and a comparable amount of storage, where p is the largest prime factor of $q-1$. Pohlig and Hellman showed that if

$$q - 1 = \prod_{i=1}^{k} p_i^{\alpha_i}, \tag{4.3}$$

where the p_i are distinct primes and the α_i are natural numbers, and if $r_1, \cdots, r_k$ are any real numbers with $0 \le r_i \le 1$, then logarithms over GF(q) can be computed in

$$\mathcal{O}\left(\sum_{i=1}^{k} \left(\log q + p_i^{1-r_i}\left(1 + \log p_i^{r_i}\right)\right)\right) \tag{4.4}$$

field operations, using

$$\mathcal{O}\left(\log q \sum_{i=1}^{k} \left(1 + p_i^{r_i}\right)\right) \tag{4.5}$$

bits of memory, provided that a precomputation requiring

$$\mathcal{O}\left(\sum_{i=1}^{k} p_i^{r_i} \log p_i^{r_i} + \log q\right) \tag{4.6}$$

field operations is performed first. This algorithm is very efficient if q is "smooth", i.e., all the prime factors of $q-1$ are small. We shall give a brief description of the algorithm as follows:

Algorithm 4.3.1 (Silver–Pohlig–Hellman Algorithm). This algorithm computes the discrete logarithm $x = \log_a b \bmod q$:

[1] Factor $q-1$ into its prime factorization form:

$$q - 1 = \prod_{i=1}^{k} p_1^{\alpha_1} p_2^{\alpha_2} \cdots p_k^{\alpha_k}.$$

[2] Precompute the table $r_{p_i,j}$ for a given field:

$$r_{p_i,j} = a^{j(q-1)/p_i} \bmod q, \quad 0 \leq j < p_i. \tag{4.7}$$

This only needs to be done once for any given field.

[3] Compute the discrete logarithm of b to the base a modulo q, i.e., compute $x = \log_a b \bmod q$:

[3-1] Use an idea similar to that in the baby-step giant-step algorithm to find the individual discrete logarithms $x \bmod p_i^{\alpha_i}$: To compute $x \bmod p_i^{\alpha_i}$, we consider the representation of this number to the base p_i:

$$x \bmod p_i^{\alpha_i} = x_0 + x_1 p_i + \cdots + x_{\alpha_i - 1} p_i^{\alpha_i - 1}, \tag{4.8}$$

where $0 \leq x_n < p_i - 1$.

(a) To find x_0, we compute $b^{(q-1)/p_i}$ which equals $r_{p_i,j}$ for some j, and set $x_0 = j$ for which

$$b^{(q-1)/p_i} \bmod q = r_{p_i,j}.$$

This is possible because

$$b^{(q-1)/p_i} \equiv a^{x(q-1)/p} \equiv a^{x_0(q-1)/p} \bmod q = r_{p_i,x_0}$$

(b) To find x_1, compute $b_1 = ba^{-x_0}$. If

$$b_1^{(q-1)/p_i^2} \bmod q = r_{p_i,j}$$

then set $x_1 = j$. This is possible because

$b_1^{(q-1)/p_i^2} \equiv a^{(x-x_0)(q-1)/p_i^2} \equiv a^{(x_1+x_2p_i+\cdots)(q-1)/p_i}$
$\equiv a^{x_1(q-1)/p} \bmod q = r_{p_i,x_1}$

(c) To obtain x_2, consider the number $b_2 = ba^{-x_0-x_1p_i}$ and compute

$$b_2^{(q-1)/p_i^3} \bmod q.$$

The procedure is carried on inductively to find all $x_0, x_1, \cdots, x_{\alpha_i - 1}$.

[3-2] Use the Chinese Remainder Theorem to find the unique value of x from the congruences $x \bmod p_i^{\alpha_i}$.

We now give an example of how the above algorithm works:

Example 4.3.1. Suppose we wish to compute the discrete logarithm $x = \log_2 62 \bmod 181$. Now we have $a = 2$, $b = 62$ and $q = 181$ (2 is a generator of $\mathbb{F}_{181}^*$). We follow the computation steps described in the above algorithm:

[1] Factor $q - 1$ into its prime factorization form:

$$180 = 2^2 \cdot 3^2 \cdot 5.$$

[2] Use the following formula to precompute the table $r_{p_i,j}$ for the given field $\mathbb{F}_{181}^*$:

$$r_{p_i,j} = a^{j(q-1)/p_i} \bmod q, \quad 0 \le j < p_i.$$

This only needs to be done once for this field.

(a) Compute $r_{p_1,j} = a^{j(q-1)/p_1} \bmod q = 2^{90j} \bmod 181$ for $0 \le j < p_1 = 2$:

$$r_{2,0} = 2^{90 \cdot 0} \bmod 181 = 1,$$
$$r_{2,1} = 2^{90 \cdot 1} \bmod 181 = 180.$$

(b) Compute $r_{p_2,j} = a^{j(q-1)/p_2} \bmod q = 2^{60j} \bmod 181$ for $0 \le j < p_2 = 3$:

$$r_{3,0} = 2^{60 \cdot 0} \bmod 181 = 1,$$
$$r_{3,1} = 2^{60 \cdot 1} \bmod 181 = 48,$$
$$r_{3,2} = 2^{60 \cdot 2} \bmod 181 = 132.$$

(c) Compute $r_{p_3,j} = a^{j(q-1)/p_3} \bmod q = 2^{36j} \bmod 181$ for $0 \le j < p_3 = 5$:

$$r_{5,0} = 2^{36 \cdot 0} \bmod 181 = 1,$$
$$r_{5,1} = 2^{36 \cdot 1} \bmod 181 = 59,$$
$$r_{5,2} = 2^{36 \cdot 2} \bmod 181 = 42,$$
$$r_{5,3} = 2^{36 \cdot 3} \bmod 181 = 125,$$
$$r_{5,4} = 2^{36 \cdot 4} \bmod 181 = 135.$$

Construct the $r_{p_i,j}$ table as follows:

p_i	j				
	0	1	2	3	4
2	1	180			
3	1	48	132		
5	1	59	42	125	135

This table is manageable if all p_i are small.

[3] Compute the discrete logarithm of 62 to the base 2 modulo 181, that is, compute $x = \log_2 62 \bmod 181$. Here $a = 2$ and $b = 62$:

[3-1] Find the individual discrete logarithms $x \bmod p_i^{\alpha_i}$ using

$$x \bmod p_i^{\alpha_i} = x_0 + x_1 p_i + \cdots + x_{\alpha_i - 1} p_i^{\alpha_i - 1}, \quad 0 \le x_n < p_i - 1.$$

(a-1) Find the discrete logarithms $x \bmod p_1^{\alpha_1}$, i.e., $x \bmod 2^2$:

$$x \bmod 181 \Longleftrightarrow x \bmod 2^2 = x_0 + 2x_1.$$

(i) To find x_0, we compute

$$b^{(q-1)/p_1} \bmod q = 62^{180/2} \bmod 181 = 1 = r_{p_1,j} = r_{2,0}$$

hence $x_0 = 0$.

(ii) To find x_1, compute first $b_1 = ba^{-x_0} = b = 62$, then compute

$$b_1^{(q-1)/p_1^2} \bmod q = 62^{180/4} \bmod 181 = 1 = r_{p_1,j} = r_{2,0}$$

hence $x_1 = 0$. So

$$x \bmod 2^2 = x_0 + 2x_1 \Longrightarrow x \bmod 4 = 0.$$

(a-2) Find the discrete logarithms $x \bmod p_2^{\alpha_2}$, that is, $x \bmod 3^2$:

$$x \bmod 181 \Longleftrightarrow x \bmod 3^2 = x_0 + 2x_1.$$

(i) To find x_0, we compute

$$b^{(q-1)/p_2} \bmod q = 62^{180/3} \bmod 181 = 48 = r_{p_2,j} = r_{3,1}$$

hence $x_0 = 1$.

(ii) To find x_1, compute first $b_1 = ba^{-x_0} = 62 \cdot 2^{-1} = 31$, then compute

$$b_1^{(q-1)/p_2^2} \bmod q = 31^{180/3^2} \bmod 181 = 1 = r_{p_2,j} = r_{3,0}$$

hence $x_1 = 0$. So

$$x \bmod 3^2 = x_0 + 2x_1 \Longrightarrow x \bmod 9 = 1.$$

(a-3) Find the discrete logarithms $x \bmod p_3^{\alpha_3}$, that is, $x \bmod 5^1$:

$$x \bmod 181 \Longleftrightarrow x \bmod 5^1 = x_0.$$

To find x_0, we compute

$$b^{(q-1)/p_3} \bmod q = 62^{180/5} \bmod 181 = 1 = r_{p_3,j} = r_{5,0}$$

hence $x_0 = 0$. So we conclude that

$$x \bmod 5 = x_0 \Longrightarrow x \bmod 5 = 0.$$

[3-2] Find the x in

$$x \bmod 181$$

such that

$$\begin{cases} x \bmod 4 = 0, \\ x \bmod 9 = 1, \\ x \bmod 5 = 0. \end{cases}$$

To do this, we just use the Chinese Remainder Theorem to solve the following system of congruences:

$$\begin{cases} x \equiv 0 \pmod 4 \\ x \equiv 1 \pmod 9 \\ x \equiv 0 \pmod 5 \end{cases}$$

The unique value of x for this system of congruences is $x = 100$. (This can be easily done by using, for example, the Maple function `chrem([0, 1, 0], [4,9, 5])`.) So the value of x in the congruence $x \bmod 181$ is 100. Hence $x = \log_2 62 = 100$.

4.4 Index Calculus Attacks

In this section we discuss some subexponential-time index calculus algorithms for DLP.

Algorithm 4.4.1 (Index calculus for DLP). This algorithm tries to find an integer k such that

$$k \equiv \log_\beta \alpha \pmod p \quad \text{or} \quad \alpha \equiv \beta^k \pmod p.$$

[1] Precomputation

[1-1] (Choose Factor Base) Select a factor base Γ, consisting of the first m prime numbers,

$$\Gamma = \{p_1, p_2, \cdots, p_m\},$$

with $p_m \le B$, the bound of the factor base.

[1-2] (Compute $\beta^e \bmod p$) Randomly choose a set of exponent $e \le p-2$, compute $\beta^e \bmod p$, and factor it as a product of prime powers.

[1-3] (Smoothness) Collect only those relations $\beta^e \bmod p$ that are smooth with respect to B. That is,

$$\beta^e \bmod p = \prod_{i=1}^{m} {p_i}^{e_i}, e_i \geq 0. \tag{4.9}$$

When such relations exist, get

$$e \equiv \sum_{j=1}^{m} e_j \log_\beta p_j \pmod{p-1}. \tag{4.10}$$

[1-4] (Repeat) Repeat [1-3] to find at least m such e in order to find m relations as in (4.10) and solve $\log_\beta p_j$ for $j = 1, 2, \cdots, m$.

[2] Compute $k \equiv \log_\beta \alpha \pmod{p}$

[2-1] For each e in (4.10), determine the value of $\log_\beta p_j$ for $j = 1, 2, \cdots, m$ by solving the m modular linear equations with unknown $\log_\beta p_j$.

[2-2] (Compute $\alpha\beta^r \bmod p$) Randomly choose exponent $r \leq p - 2$ and compute $\alpha\beta^r \bmod p$.

[2-3] (Factor $\alpha\beta^r \bmod p$ over Γ)

$$\alpha\beta^r \bmod p = \prod_{j=1}^{m} {p_j}^{r_i}, r_j \geq 0. \tag{4.11}$$

If (4.11) is unsuccessful, go back to [2-2]. If it is successful, then

$$\log_\beta \alpha \equiv -r + \sum_{j=1}^{m} r_j \log_\beta p_j. \tag{4.12}$$

Example 4.4.1 (Index calculus for DLP). Find

$$x \equiv \log_{22} 4 \pmod{3361}$$

such that

$$4 \equiv 22^x \pmod{3361}.$$

[1] Precomputation

[1-1] (Choose Factor Base) Select a factor base Γ, consisting of the first 4 prime numbers,

$$\Gamma = \{2, 3, 5, 7\},$$

with $p_4 \leq 7$, the bound of the factor base.

[1-2] (Compute $22^e \bmod 3361$) Randomly choose a set of exponent $e \leq 3359$, compute $22^e \bmod 3361$, and factor it as a product of prime powers.

$$22^{48} \equiv 2^5 \cdot 3^2 \pmod{3361}$$
$$22^{100} \equiv 2^6 \cdot 7 \pmod{3361}$$
$$22^{186} \equiv 2^9 \cdot 5 \pmod{3361}$$
$$22^{2986} \equiv 2^3 \cdot 3 \cdot 5^2 \pmod{3361}$$

[1-3] (Smoothness) The above four relations are smooth with respect to $B = 7$. Thus

$$48 \equiv 5\log_{22} 2 + 2\log_{22} 3 \pmod{3360}$$
$$100 \equiv 6\log_{22} 2 + \log_{22} 7 \pmod{3360}$$
$$186 \equiv 9\log_{22} 2 + \log_{22} 5 \pmod{3360}$$
$$2986 \equiv 3\log_{22} 2 + \log_{22} 3 + 2\log_{22} 5 \pmod{3360}$$

[2] Compute $k \equiv \log_\beta \alpha \pmod{p}$

[2-1] Compute

$$\log_{22} 2 = 1100$$
$$\log_{22} 3 = 2314$$
$$\log_{22} 5 = 366$$
$$\log_{22} 7 = 220$$

[2-2] (Compute $4 \cdot 22^r \bmod p$) Randomly choose exponent $r = 754 \leq 3659$ and compute $4 \cdot 22^{3659} \bmod 3361$.

[2-3] (Factor $4 \cdot 22^{754} \bmod 3361$ over Γ)

$$4 \cdot 22^{754} \equiv 2 \cdot 3^2 \cdot 5 \cdot 7 \pmod{3361}. \quad (4.13)$$

Thus,

$$\log_{22} 4 \equiv -754 + \log_{22} 2 + 2\log_{22} 3 + \log_{22} 5 + \log_{22} 7 \equiv 2200. \quad (4.14)$$

That is,

$$22^{2200} \equiv 4 \pmod{3361}.$$

Example 4.4.2. Find $k \equiv \log_{11} 7 \pmod{29}$ such that $\beta^k \equiv 11 \pmod{29}$.

[1] (Factor Base) Let the factor base $\Gamma = \{2, 3, 5\}$.

[2] (Compute and Factor $\beta^e \bmod p$) Randomly choose $e < p$, compute and factor $\beta^e \bmod p = 11^e \bmod 29$ as follows:

(1) $11^2 \equiv 5 \pmod{29}$	(success)
(2) $11^3 \equiv 2 \cdot 13 \pmod{29}$	(fail)
(3) $11^5 \equiv 2 \cdot 7 \pmod{29}$	(fail)
(4) $11^6 \equiv 3^2 \pmod{29}$	(success)
(5) $11^7 \equiv 2^3 \cdot 3 \pmod{29}$	(success)
(6) $11^9 \equiv 2 \cdot 7 \pmod{29}$	(success)

[3] (Solve the systems of congruences for the quantities $\log_\beta p_i$)

$$\begin{aligned}
&(1)\ \log_{11} 5 \equiv 2 \pmod{28} \\
&(4)\ \log_{11} 3 \equiv 3 \pmod{28} \\
&(6)\ \log_{11} 2 \equiv 9 \pmod{28} \\
&(5)\ 2 \cdot \log_{11} 2 + \log_{11} 3 \equiv 7 \pmod{28} \\
&\quad\ \log_{11} 3 \equiv 17 \pmod{28}
\end{aligned}$$

[4] (Compute and Factor $\alpha\beta^e \bmod p$) Randomly choose $e < p$, compute and factor $\alpha\beta^e \bmod p = 7 \cdot 11^e \bmod 29$ as follows:

$$\begin{array}{ll}
7 \cdot 11 \equiv 19 \pmod{29} & \text{(fail)} \\
7 \cdot 11^2 \equiv 2 \cdot 3 \pmod{29} & \text{(success)}
\end{array}$$

Thus

$$\log_{11} 7 \equiv \log_{11} 2 + \log_{11} 3 - 2 \equiv 24 \pmod{28}.$$

This is true since

$$11^{24} \equiv 7 \pmod{29}.$$

In what follows, we shall briefly introduce two subexponential algorithms for taking discrete logarithms $x = \log_a b \bmod q$, namely Adleman's index calculus algorithm and Gordon's number field sieve method.

In 1979, Adleman proposed a general purpose, subexponential algorithm for taking discrete logarithms, called the *index calculus method*, with the following expected running time:

$$\mathcal{O}\left(\exp\left(c\sqrt{\log q \log\log q}\right)\right).$$

His algorithm can be briefly described as follows:

Algorithm 4.4.2 (Adleman's Index calculus). This algorithm tries to compute the discrete logarithm $x = \log_a b \bmod q$ with input a, b, q, where a and b are generators and q a prime:

[1] Factor $q-1$ into its prime factorization form:

$$q - 1 = p_1^{\alpha_1} p_2^{\alpha_2} \cdots p_k^{\alpha_k}$$

[2] For each $p_k^{\alpha_k} \mid n$ carry out the following steps until m_l is obtained:

- (Guessing and checking) Find r_i, s_i such that $a^{r_i} \bmod q$ and $ba^{s_i} \bmod q$ are smooth with respect to the bound $2^{(\log q \log\log q)^{1/2}}$.
- (Using Gaussian elimination) Check if over the finite field $\mathbb{Z}_{p_l^{\alpha_l}}$, $ba^{s_i} \bmod q$ is dependent on

$$\{a^{r_1} \bmod q, \cdots, a^{r_i} \bmod q\}.$$

If yes, calculate β_j's such that

$$ba^{s_i} \bmod q \equiv \left(\sum_{j=1}^{i} \beta_j a^{r_i} \bmod q \right) \bmod p_l^{\alpha_l}$$

then

$$m_l = \left(\sum_{j=1}^{i} \beta_j r_j \right) \bmod p_l^{\alpha_l} \; - \; s_i$$

[3] (Using the Chinese Remainder Theorem) Calculate and output x such that

$$x \equiv m_l \pmod{p_l^{\alpha_l}}, \quad l = 1, 2, \cdots k.$$

Note that the above algorithm can also be easily generalized to the case where q is not a prime, or a or b are not generators.

For more than ten years after its invention, the above algorithm and its variants were the fastest algorithms for computing discrete logarithms. But the situation changed when Gordon in 1993 proposed an algorithm for computing discrete logarithms in GF(p). Gordon's algorithm is based on the Number Field Sieve (NFS) for integer factorization, with the heuristic expected running time

$$\mathcal{O}\left(\exp\left(c(\log p)^{1/3}(\log\log p)^{2/3}\right)\right),$$

the same as that used in factoring. The algorithm can be briefly described as follows:

Algorithm 4.4.3 (Gordon's NFS). This algorithm computes the discrete logarithm x such that $a^x \equiv b \pmod{p}$ with input a, b, p, where a and b are generators and p is prime:

[1] (Precomputation): Find the discrete logarithms of a factor base of small rational primes, which must only be done once for a given p.

[2] (Compute individual logarithms): Find the logarithm for each $b \in \mathbb{F}_p$ by finding the logarithms of a number of "medium-sized" primes.

[3] (Compute the final logarithm): Combine all the individual logarithms (by using the Chinese Remainder Theorem) to find the logarithm of b.

Interested readers are referred to Gordon's paper [127] for more detailed information. Note also that Gordon, with co-author McCurley [126], discussed some implementation issues of massively parallel computations of discrete logarithms over $\mathbb{F}_{2^n}$.

4.5 Xedni Calculus Attack

The xedni calculus was first proposed by Joseph Silverman in 1998 [293], and analyzed in [156] [175] and [295]. It is called *xedni calculus* because it "stands index calculus on its head". The xedni calculus is a new method that *might* be used to solve the ECDLP, although it has not yet been tested in practice. It can be described as follows [293]:

[1] Choose points in $E(\mathbb{F}_p)$ and lift them to points in $\mathbb{Z}^2$.

[2] Choose a curve $\mathcal{E}(\mathbb{Q})$ containing the lift points; use Mestre's method [210] (in reverse) to make rank $\mathcal{E}(\mathbb{Q})$ small.

Whilst the index calculus works in reverse:

[1] Lift $E/\mathbb{F}_p$ to $\mathcal{E}(\mathbb{Q})$; use Mestre's method to make rank $\mathcal{E}(\mathbb{Q})$ large.

[2] Choose points in $E(\mathbb{F}_p)$ and try to lift them to points in $\mathcal{E}(\mathbb{Q})$.

A brief description of the xedni algorithm is as follows (a completed description and justification of the algorithm can be found in [293]).

Algorithm 4.5.1 (Xedni calculus for the ECDLP). Let $\mathbb{F}_p$ be a finite field with p elements (p prime), $E/\mathbb{F}_p$ an elliptic curve over $\mathbb{F}_p$, say, given by

$$E: \quad y^2 + a_{p,1}xy + a_{p,3}y = x^3 + a_{p,2}x^2 + a_{p,4}x + a_{p,6}.$$

N_p the number of points in $E(\mathbb{F}_p)$, S and T the two points in $E(\mathbb{F}_p)$. This algorithm tries to find an integer k

$$k = \log_T S$$

such that

$$S = kT \quad \text{in } E(\mathbb{F}_p).$$

[1] Fix an integer $4 \leq r \leq 9$ and an integer M which is a product of small primes.

[2] Choose r points:

$$P_{M,i} = [x_{M,i}, y_{M,i}, z_{M,i}], \quad 1 \leq i \leq r \tag{4.15}$$

having integer coefficients and satisfying
- the first 4 points are $[1,0,0]$, $[0,1,0]$, $[0,0,1]$ and $[1,1,1]$.
- For every prime $l \mid M$, the matrix $\mathbf{B}(P_{M,1}, \cdots, P_{M,r})$ has maximal rank modulo l.

Further choose coefficients $u_{M,1}, \cdots, u_{M,10}$ such that the points $P_{M,1}, \cdots, P_{M,r}$ satisfy the congruence:

$$\begin{aligned} u_{M,1}x^3 + u_{M,2}x^2y + u_{M,3}xy^2 + u_{M,4}y^3 + u_{M,5}x^2z + u_{M,6}xyz + u_{M,7}y^2z \\ + u_{M,8}xz^2 + u_{M,9}yz^2 + u_{M,10}z^3 \equiv 0 \pmod{M}. \end{aligned} \tag{4.16}$$

[3] Choose r random pairs of integers (s_i, t_i) satisfying $1 \leq s_i, t_i < N_p$, and for each $1 \leq i \leq r$, compute the point $P_{p,i} = (x_{p,i}, y_{p,i})$ defined by

$$P_{p,i} = s_i S - t_i T \quad \text{in } E(\mathbb{F}_p). \tag{4.17}$$

[4] Make a change of variables in $\mathbb{P}^2$ of the form

$$\begin{pmatrix} X' \\ Y' \\ Z' \end{pmatrix} = \begin{pmatrix} a_{11} & a_{12} & a_{13} \\ a_{21} & a_{22} & a_{23} \\ a_{21} & a_{32} & a_{33} \end{pmatrix} \begin{pmatrix} X \\ Y \\ Z \end{pmatrix} \tag{4.18}$$

so that the first four points become

$$P_{p,1} = [1,0,0], \; P_{p,2} = [0,1,0], \; P_{p,3} = [0,0,1], \; P_{p,4} = [1,1,1].$$

The equation for E will then have the form:

$$\begin{aligned} &u_{p,1}x^3 + u_{p,2}x^2y + u_{p,3}xy^2 + u_{p,4}y^3 + u_{p,5}x^2z + u_{p,6}xyz \\ &+u_{p,7}y^2z + u_{p,8}xz^2 + u_{p,9}yz^2 + u_{p,10}z^3 = 0. \end{aligned} \tag{4.19}$$

[5] Use the Chinese Remainder Theorem to find integers $u'_1, \cdots, u'_{10}$ satisfying

$$u'_i \equiv u_{p,i} \pmod{p} \text{ and } u'_i \equiv u_{M,i} \pmod{M} \text{ for all } 1 \leq i \leq 10. \tag{4.20}$$

[6] Lift the chosen points to $\mathbb{P}^2(\mathbb{Q})$. That is, choose points

$$P_i = [x_i, y_i, z_i], \quad 1 \leq i \leq r, \tag{4.21}$$

with integer coordinates satisfying

$$P_i \equiv P_{p,i} \pmod{p} \text{ and } P_i \equiv P_{M,i} \pmod{M} \text{ for all } 1 \leq i \leq r. \tag{4.22}$$

In particular, take $P_1 = [1,0,0], P_2 = [0,1,0], P_3 = [0,0,1], P_4 = [1,1,1]$.

[7] Let $\mathbf{B} = \mathbf{B}(P_1, \cdots, P_r)$ be the matrix of cubic monomials defined earlier. Consider the system of linear equations:

$$\mathbf{Bu} = 0. \tag{4.23}$$

Find a small integer solution $\mathbf{u} = [u_1, \cdots, u_{10}]$ to (4.23) which has the additional property

$$\mathbf{u} \equiv [u'_1, \cdots, u'_{10}] \pmod{M_p}, \tag{4.24}$$

where $u'_1, \cdots, u'_{10}$ are the coefficients computed in Step [5]. Let $C_{\mathbf{u}}$ denote the associated cubic curve:

$$\begin{aligned} C_{\mathbf{u}} :\; &u_1x^3 + u_2x^2y + u_3xy^2 + u_4y^3 + u_5x^2z + u_6xyz \\ &+u_7y^2z + u_8xz^2 + u_9yz^2 + u_{10}z^3 = 0. \end{aligned} \tag{4.25}$$

[8] Make a change of coordinates to put $C_{\mathbf{u}}$ into standard minimal Weierstrass form with the point $P_1 = [1, 0, 0]$ the point at infinity, $\mathcal{O}$. Write the resulting equation as

$$E_{\mathbf{u}} : y^2 + a_1xy + a_3y = x^3 + a_2x^2 + a_4x + a_6 \quad (4.26)$$

with $a_1, \cdots, a_6 \in \mathbb{Z}$, and let $Q_1, Q_2, \cdots, Q_r$ denote the images of $P_1, P_2, \cdots, P_r$ under this change of coordinates (so in particular, $Q_1 = \mathcal{O}$). Let $c_4(\mathbf{u})$, $c_6(\mathbf{u})$, and $\Delta(\mathbf{u})$ be the usual quantities in [291] associated to the equation (4.26).

[9] Check if the points $Q_1, Q_2, \cdots, Q_r \in E_{\mathbf{u}}(\mathbb{Q})$ are independent. If they are, return to Step [2] or [3]. Otherwise compute a relation of dependence

$$n_2Q_2 + n_3Q_3 + \cdots + n_rQ_r = \mathcal{O}, \quad (4.27)$$

set

$$n_1 = -n_2 - n_3 - \cdots - n_r, \quad (4.28)$$

and continue with the next step.

[10] Compute

$$s = \sum_{i=1}^{r} n_i s_i \quad \text{and} \quad t = \sum_{i=1}^{r} n_i t_i. \quad (4.29)$$

If $\gcd(s, N_p) > 1$, return to Step [2] or [3]. Otherwise compute an inverse $ss' \equiv 1 \pmod{N_p}$. Then

$$\log_T S \equiv s't \pmod{N_p}, \quad (4.30)$$

and the ECDLP is solved.

As can be seen, the basic idea in the above algorithm is that we first choose points $P_1, P_2 \cdots, P_r$ in $E(\mathbb{F}_p)$ and lift them to points $Q_1, Q_2 \cdots, Q_r$ having integer coordinates, then we choose an elliptic curve $\mathcal{E}(\mathbb{Q})$ that goes through the points $Q_1, Q_2 \cdots, Q_r$, finally, check if the points $Q_1, Q_2 \cdots, Q_r$ are *dependent*. If they are, the ECDLP is almost solved. Thus, the goal of the xedni calculus is to find an instance where an elliptic curve has *smaller* than expected rank. Unfortunately, a set of points $Q_1, Q_2 \cdots, Q_r$ as constructed above will usually be *independent*. So, it will not work. To make it work, a congruence method, due to Mestre [210], is used *in reverse* to produce the lifted curve $\mathcal{E}$ having smaller than expected rank[1]. Again unfortunately, Mestre's method is based on some deep ideas and unproved conjectures in analytic number theory and arithmetic algebraic geometry, it is not possible for us at present to give even a rough estimate of the algorithm's running time. So, in reality we know nothing about the complexity of the xedni calculus. We also do not know if the xedni calculus will be practically useful; it may

[1] Mestre's original method is to produce elliptic curves of large rank.

be completely useless from a practical point of view. Much needs to be done before we can have a better understanding of the xedni calculus. As we know the index calculus is probabilistic, subexponential-time algorithm applicable for both the integer factorization problem (IFP) and the finite field discrete logarithm problem (DLP). However, there is no known subexponential-time algorithm for the elliptic curve discrete logarithm (ECDLP); whether or not there exists a subexponential-time algorithm for ECDLP is an important open problem in mathematics, cryptography and computer science.

In November 1997, Certicom, a computer security company in Waterloo, Canada, introduced the Elliptic Curve Cryptosystem (ECC) Challenge, developed to increase industry understanding and appreciation for the difficulty of ECDLP and to encourage and stimulate further research in the security analysis of ECC. The challenge is to compute the ECC private keys from the given list of ECC public keys and associated system parameters. It is the type of problem facing an adversary who wishes to attack ECC. These problems are defined on curves either over $\mathbb{F}_{2^m}$ or over $\mathbb{F}_p$ with p prime (see Table 4.1 and Table 4.2. Also there are three levels of difficulties associated to the curves: exercise level (with bits less than 109), rather easy level (with bits in 109-131), and very hard level (with bits in 163-359).

Curve	Field size (in bits)	Estimated number of machine days	Prize in US dollars	Status
ECC2-79	79	352	HAC & Maple V	Dec 1997
ECC2-89	89	11278	HAC & Maple V	Feb 1998
ECC2K-95	97	8637	5,000	May 1998
ECC2-97	97	180448	5,000	Sept 1999
ECC2K-108	109	1.3 × 106	10,000	April 2000
ECC2-109	109	2.1 × 107	10,000	April 2004
ECC2K-130	131	2.7 × 109	20,000	?
ECC2-131	131	6.6 × 1010	20,000	?
ECC2-163	163	2.9 × 1015	30,000	?
ECC2K-163	163	4.6 × 1014	30,000	?
ECC2-191	191	1.4 × 1020	40,000	?
ECC2-238	239	3.0 × 1027	50,000	?
ECC2K-238	239	1.3 × 1026	50,000	?
ECC2-353	359	1.4 × 1045	100,000	?
ECC2K-358	359	2.8 × 1044	100,000	?

Table 4.1. Elliptic Curves over $\mathbb{F}_{2^m}$

Now let us take the random curve ECC2-79 over $\mathbb{F}_{2^{79}}$ in the Challenge as an example:

```
m = 79
f = x79 +x9 + 1
seedE = D3E5D53A 4D696E67 68756151 757779B8 379AC409
```

Curve	Field size (in bits)	Estimated number of machine days	Prize in US dollars	Status
ECCp-79	79	146	HAC & Maple V	Dec 1997
ECCp-89	89	4360	HAC & Maple V	Jan 1998
ECCp-97	97	71982	5,000	March 1998
ECCp-109	109	9 × 107	10,000	Nov 2002
ECCp-131	131	2.3 × 1010	20,000	?
ECCp-163	163	2.3 × 1015	30,000	?
ECCp-191	191	4.8 × 1019	40,000	?
ECCp-239	239	1.4 × 1027	50,000	?
ECC2p-359	359	3.7 × 1045	100,000	?

Table 4.2. Elliptic Curves over $\mathbb{F}_p$

```
a = 4A2E 38A8F66D 7F4C385F
b = 2C0B B31C6BEC C03D68A7
seedP = 50CBF1D9 5CA94D69 6E676875 615175F1 6A36A3B1
U_x = 5674 7BC3DDBF F399EF4B
U_y = 49FF 222B2065 008D3C2C
P_x = 30CB 127B63E4 2792F10F
P_y = 547B 2C88266B B04F713B
h = 02
n = 4000 00000045 31A2562B
seedQ = 4EC95934 D696E676 87561517 56D5D2A8 FCD02E68
V_x = 5B6E 06AFB573 C1CF2BF5
V_y = 69F9 74972600 1DB5B8D9
Q_x = 0020 2A9F0350 14497325
Q_y = 5175 A6485955 2F97C129
```

A brief explanation of the curve (format and notation, etc.) is as follows (more detailed explanation on all other types of curves in the Challenger can be found in the following website):

`http://www.certicom.com/index.php?action=ecc,ecc_challenge`

[1] $m = 79$: m is the order of the finite field $\mathbb{F}_{2^{79}}$.

[2] $f(x) = x^{79} + x^9 + 1$: The reduction polynomial used to represent the elements in $\mathbb{F}_{2^{79}}$.

[3] seedE the seed used to generate the parameters (a, b).

[4] (a, b): the field elements used to define the elliptic curve

$$E : y^2 = x^3 + ax + b.$$

[5] seedP: the seed used to generate the base point P on E.

[6] (x_p, y_p): the x- and y-coordinates of the based point P.

[7] n: the order of the point P, with n prime.

[8] h: co-factor h (the number of points in $E(\mathbb{F}_2^{79})$ divided by n).

[9] seedQ: the seed used to generate the base point Q on E.

[10] (x_q, y_q): the x- and y-coordinates of the based point Q.

[11] Integers are represented in hexadecimal.

[12] Field elements are represented in hexadecimal.

[13] Seeds are used for generating random elliptic curves and random points on curves.

For this particular example, the answer is

$$\log_p Q = 3AA068A09FIED21E2582$$

The result was obtained by using parallelized Pollard ρ-method with 1737410165381 iterations and 1767 distinguished points.

4.6 Chapter Notes and Further Reading

The problem of computing discrete logarithms (DLP) is fundamental in computational number theory, computational algebra, and of great importance in public-key cryptography and information security. Remarkably enough, the DLP problem has much in common with the IFP problem. First of all, they are both computationally intractable, with almost the same level of intractability. As a rule of thumb, the complexity of factoring N with β digits is almost the same as that of taking discrete logarithms modulo a prime p with $\beta - 10$ digits. Secondly, the methods used for IFP are always applicable to DLP, and sometimes even applicable to ECDLP (see Table 4.3 for more information): Finally, IFP, DLP and ECDLP are all useful in the design and

IFP	DLP	ECDLP
Trial Divisions	Baby-Step Giant-Step	
Pollard's ρ-method	Pollard's λ-method	
CFRAC/MPQS	Index Calculus	
NFS	NFS	
Xedni Calculus	Xedni Calculus	Xedni Calculus
Quantum Algorithms	Quantum Algorithms	Quantum Algorithms

Table 4.3. Algorithms for IFP, DLP and ECDLP

implementation of many cryptographic systems, the security of such systems are based on the intractability of these problems.

Although efficient algorithms still have not been found for IFP, DLP, and ECDLP, a great progress has been made in recent years. Readers may consult [86], [201], [232], [233], [249], [276] for more information of DLP, and [292], [295], [293], [336], and [337] for ECDLP. Again, quantum computers and quantum algorithms are very well suited for IFP, DLP and ECDLP, but the (new) problem is that a practically useful quantum computer is not available at present and may never be able to be built. Nevertheless, there are quantum algorithms for IFP, DLP and ECDLP, which are the subject matters of our next chapter.

5. Quantum Computing Attacks

The quest for efficiency in computational methods yields not only fast algorithms, but also insights that leads to elegant, simple, and general problem-solving methods.

ROBERT E. TARJAN
The 1986 Turing Award Recipient

5.1 Introduction

In the RSA cryptosystem, it is assumed that the cryptanalyst, Eve

(1) knows the public-key $\{e, N\}$ with $N = pq$ and also the ciphertext C,

(2) does not know any one piece of the trap-door information $\{p, q, \phi(N), d\}$,

(3) wants to know $\{M\}$.

That is,

$$\{e, N, C \equiv M^e \pmod{N}\} \xrightarrow{\text{Eve wants to find}} \{M\}.$$

Obviously, there are several ways to recover M from C (i.e., to break the RSA system):

(1) Factor N to get $\{p, q\}$ so as to compute

$$M \equiv C^{1/e \pmod{(p-1)(q-1)}} \pmod{N}.$$

(2) find $\phi(N)$ so as to compute

$$M \equiv C^{1/e \pmod{\phi(N)}} \pmod{N}.$$

(3) Find order(a, N), the order of a random integer $a \in [2, N-2]$ modulo N, then try to find

$$\{p, q\} = \gcd(a^{r/2} \pm 1, N) \text{ and } M \equiv C^{1/e \pmod{(p-1)(q-1)}} \pmod{N}.$$

(4) Find order(C, N), the order of C modulo N, so as to compute

$$M \equiv C^{1/e \pmod{\text{order}(C,N)}} \pmod{N}.$$

(5) Compute $\log_C M \pmod{N}$, the discrete logarithm M to the base C modulo N in order to find

$$M \equiv C^{\log_C M \pmod{N}} \pmod{N}$$

Remarkably enough, all the above five operations can be efficiently performed by quantum algorithms on quantum computers, and as discussed in the previous two chapters, *cannot* be efficiently performed by classical algorithms on electronic computers. Incidently, all these operations are intimately connected to the order-finding problem which can be done in polynomial-time on a quantum computer. More specifically, using the quantum order finding algorithm (see Figure 5.1), RSA can be cracked completely in polynomial-

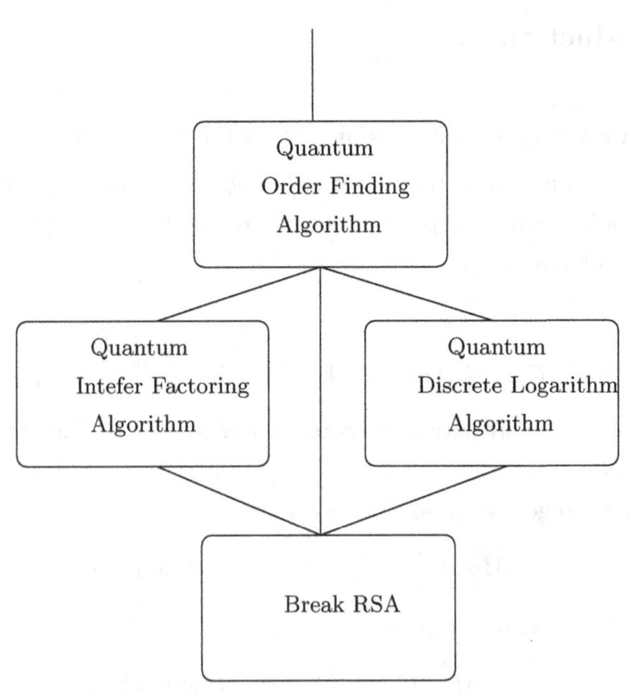

Figure 5.1. Quantum Attacks on RSA

time on a quantum computer, and the IFP, DLP, ECDLP problems can also be performed in polynomial-time on a quantum computer. In this chapter, we first introduce a quantum order find algorithm and a quantum ordering based attack, then in the two sections that follows, we discuss the quantum algorithms for integer factoring and quantum discrete logarithms and their attacks to RSA.

5.2 Order Finding Problem

Definition 5.2.1. Let $G = \mathbb{Z}_N^*$ be a finite multiplicative group, and $x \in G$ a randomly chosen integer (element). Then order of x in G, or order of an element a modulo N, some times denoted by order(x, N), is the smallest positive integer r such that

$$x^r \equiv 1 \pmod{N}.$$

Example 5.2.1. Let $5 \in \mathbb{Z}_{104}^*$. Then order$(5, 104) = 4$, since 4 is the smallest positive integer satisfying

$$5^4 \equiv 1 \pmod{104}.$$

Theorem 5.2.1. Let G be a finite group and suppose that $x \in G$ has finite order r. If $x^k = 1$, then $r \mid k$.

Example 5.2.2. Let $5 \in \mathbb{Z}_{104}^*$. As $5^{24} \equiv 1 \pmod{104}$, so, $4 \mid 24$.

Definition 5.2.2. Let G be a finite group, then the number of elements in G, denoted by $|G|$, is called the *order* of G.

Example 5.2.3. Let $G = \mathbb{Z}_{104}^*$. Then there are 48 elements in G that are relatively prime to 104 (two numbers a and b are relatively prime if $\gcd(a, b) = 1$), namely;

1, 3, 5, 7, 9, 11, 15, 17, 19, 21, 23, 25, 27, 29, 31, 33, 35, 37, 41, 43
45, 47, 49, 51, 53, 55, 57, 59, 61, 63, 67, 69, 71, 73, 75, 77, 79, 81
83, 85, 87, 89, 93, 95, 97, 99, 101, 103

Thus, $|G| = 48$. That is, the order of the group G is 48.

Theorem 5.2.2 (Lagrange). Let G be a finite group. Then the order of an element $x \in G$ divides the order of the group G.

Example 5.2.4. Let $G = \mathbb{Z}_{104}^*$. Then the order of G is 48, whereas the order of the element $5 \in G$ is 4. Clearly $4 \mid 24$.

Corollary 5.2.1. If a finite group G has order r, then $x^r = 1$ for all $x \in G$.

Example 5.2.5. Let $G = \mathbb{Z}_{104}^*$ and $|G| = 48$. Then

$$\begin{aligned} 1^{48} &\equiv 1 \pmod{104} \\ 3^{48} &\equiv 1 \pmod{104} \\ 5^{48} &\equiv 1 \pmod{104} \\ 7^{48} &\equiv 1 \pmod{104} \\ &\vdots \\ 101^{48} &\equiv 1 \pmod{104} \\ 103^{48} &\equiv 1 \pmod{104}. \end{aligned}$$

Now, we are in a position to present our two main theorems as follows.

Theorem 5.2.3. Let C be the RSA ciphertext, and order(C, N) the order of $C \in \mathbb{Z}_N^*$. Then

$$d \equiv 1/e \ (\text{mod order}(C, N)).$$

Corollary 5.2.2. Let C be the RSA ciphertext, and order(C, N) the order of $C \in \mathbb{Z}_N^*$. Then

$$M \equiv C^{1/e \ (\text{ mod order}(C,N))} \pmod N$$

Thus, to recover the RSA M from C, it suffices to just find the order of C modulo N.

Now we return to the actual computation of the order of an element x in $G = \mathbb{Z}_N^*$. Finding the order of an element x in G is, in theory, not a problem: just keep multiplying until we get to "1", the identity element of the multiplicative group G. For example, let $N = 179359$, $x = 3 \in G$, and $G = \mathbb{Z}_{179359}^*$, such that $\gcd(3, 179359) = 1$. To find the order $r = \text{order}(3, 179359)$, we just keep multiplying until we get to "1":

$$\begin{array}{llll} 3^1 & \text{mod} & 179359 & = & 3 \\ 3^2 & \text{mod} & 179359 & = & 9 \\ 3^3 & \text{mod} & 179359 & = & 27 \\ & & & \vdots & \\ 3^{1000} & \text{mod} & 179359 & = & 31981 \\ 3^{1001} & \text{mod} & 179359 & = & 95943 \\ 3^{1002} & \text{mod} & 179359 & = & 108470 \\ & & & \vdots & \\ 3^{14716} & \text{mod} & 179359 & = & 99644 \\ 3^{14717} & \text{mod} & 179359 & = & 119573 \\ 3^{14718} & \text{mod} & 179359 & = & 1. \end{array}$$

Thus, the order r of 3 in the multiplicative group $\mathcal{G} = (\mathbb{Z}/179359\mathbb{Z})^*$ is 14718, that is, $\text{ord}_{179359}(3) = 14718$.

Example 5.2.6. Let

$$\begin{aligned} N &= 5515596313 \\ e &= 1757316971 \\ C &= 763222127 \\ r &= \text{order}(C, N) = 114905160 \end{aligned}$$

Then

$$\begin{aligned} M &\equiv C^{1/e \bmod r} \pmod N \\ &\equiv 763222127^{1/1757316971 \bmod 114905160} \pmod{5515596313} \\ &\equiv 1612050119 \end{aligned}$$

Clearly, this result is correct, since

$$\begin{aligned} M^e &\equiv 1612050119^{1757316971} \\ &\equiv 763222127 \\ &\equiv C \pmod{5515596313} \end{aligned}$$

It must be noted, however, that in practice, the above computation for finding the order of $x \in (\mathbb{Z}/N\mathbb{Z})^*$ may not work, since for an element x in a large group $\mathcal{G}$ with N having more than 200 digits, the computation of r may require more than 10^{150} multiplications. Even if these multiplications could be carried out at the rate of 1000 billion per second on a supercomputer, it would take approximately $3 \cdot 10^{80}$ years to arrive at the answer. Thus, the order finding problem is intractable on conventional digital computers. The problem is, however, tractable on quantum computers, provided that a practical quantum computer is available.

5.3 Quantum Order Finding Attack

It may be the case that, as the famous physicist Feynman mentioned, nobody understands quantum mechanics, some progress has been made in quantum mechanics, particularly in quantum computing and quantum cryptography. In this section, we present a quantum algorithm for computing the order of an element x in the multiplicative group $\mathbb{Z}_N^*$, due to Shor [287]. The main idea of Shor's algorithm is as follows. First of all, we create two quantum registers for our quantum computer: Register-1 and Register-2. Of course, we can create just one single quantum memory register partitioned into two parts. Secondly, we create in Register-1, a superposition of the integers $a = 0, 1, 2, 3, \cdots$ which will be the arguments of $f(a) = x^a \pmod N$, and load Register-2 with all zeros. Thirdly, we compute in Register-2, $f(a) = x^a \pmod N$ for each input a. (Since the values of a are kept in Register-1, this can be done reversibly). Fourthly, we perform the discrete Fourier transform on Register-1. Finally we observe both registers of the machine and find the order r that satisfies $x^r \equiv 1 \pmod N$. The following is a brief description of the quantum algorithm for the order finding problem.

Algorithm 5.3.1 (Quantum Order Finding Attack). Given RSA ciphertext C and modulus N, this attack will first find the order, r, of C in $\mathbb{Z}_N^8$, such that $C^r \equiv 1 \pmod N$, then recover the plaintext M from the ciphertext C. Assume the quantum computer has two quantum registers: Register-1 and Register-2, which hold integers in binary form.

[1] (Initialization) Find a number q, a power of 2, say 2^t, with $N^2 < q < 2N^2$.

[2] (Preparation for quantum registers) Put in the first t-qubit register, Register-1, the uniform superposition of states representing numbers $a \pmod q$, and load Register-2 with all zeros. This leaves the machine in the state $|\Psi_1\rangle$:

$$|\Psi_1\rangle = \frac{1}{\sqrt{q}} \sum_{a=0}^{q-1} |a\rangle\,|0\rangle\,.$$

(Note that the joint state of both registers are represented by $|$Register-1$\rangle$ and $|$Register-2$\rangle$). What this step does is put each qubit in Register-1 into the superposition

$$\frac{1}{\sqrt{2}}\,(|0\rangle + |1\rangle)\,.$$

[3] (Power Creation) Fill in the second t-qubit register, Register-2, with powers $C^a \pmod N$. This leaves the machine in state $|\Psi_2\rangle$:

$$|\Psi_2\rangle = \frac{1}{\sqrt{q}} \sum_{a=0}^{q-1} |a\rangle\,|C^a \pmod N\rangle\,.$$

This step can be done reversibly since all the a's were kept in Register-1.

[4] (Perform a quantum FFT) Apply FFT on Register-1. The FFT maps each state $|a\rangle$ to

$$\frac{1}{\sqrt{q}} \sum_{c=0}^{q-1} \exp(2\pi iac/q)\,|c\rangle\,.$$

That is, we apply the unitary matrix with the (a, c) entry equal to $\frac{1}{\sqrt{q}} \exp(2\pi iac/q)$. This leaves the machine in the state $|\Psi_3\rangle$:

$$|\Psi_3\rangle = \frac{1}{q}\sum_{a=0}^{q-1}\sum_{c=0}^{q-1}\exp(2\pi iac/q)\,|c\rangle\,|x^a \pmod N\rangle\,.$$

[5] (Periodicity Detection in x^a) Observe both $|c\rangle$ in Register-1 and $|x^a \pmod N\rangle$ in Register-2 of the machine, measure both arguments of this superposition, obtaining the values of $|c\rangle$ in the first argument and some $\left|x^k \pmod n\right\rangle$ as the answer for the second one ($0 < k < r$).

[6] (Extract r) Extract the required value of r. Given the pure state $|\Psi_3\rangle$, the probabilities of different results for this measurement will be given by the probability distribution:

$$\begin{aligned}
\text{Prob}(c, C^k \pmod N) &= \left|\frac{1}{q}\sum_{\substack{a=0\\ C^a\equiv a^k \pmod N}}^{q-1}\exp(2\pi iac/q)\right|^2 \\
&= \left|\frac{1}{q}\sum_{B=0}^{\lfloor (q-k-1)/r\rfloor}\exp(2\pi i(br+k)c/q)\right|^2 \\
&= \left|\frac{1}{q}\sum_{B=0}^{\lfloor (q-k-1)/r\rfloor}\exp(2\pi ib\{rc\}/q)\right|^2
\end{aligned}$$

where $\{rc\}$ is $rc \bmod N$. As shown in [287],

$$\begin{aligned}
\frac{-r}{2} \le \{rc\} \le \frac{-r}{2} &\implies \frac{-r}{2} \le rc - dq \le \frac{-r}{2}, \text{ for some } d \\
&\implies \text{Prob}(c, C^k \pmod N) > \frac{1}{3r^2}.
\end{aligned}$$

then we have

$$\left|\frac{c}{q} - \frac{d}{r}\right| \le \frac{1}{2q}.$$

Since $\frac{c}{q}$ were known, r can be obtained by the continued fraction expansion of $\frac{c}{q}$.

[7] (Code Breaking) Once the order r, $r = \text{order}(C, N)$, is found, then compute:

$$M \equiv C^{1/e \bmod r} \pmod N.$$

Hence, decodes the RSA code C.

Theorem 5.3.1. (Complexity of Quantum Order Finding Attack). Quantum order attack can find order(C, N) and recover M from C in time $\mathcal{O}((\log N)^{2+\epsilon})$.

Remark 5.3.1. This quantum attack is for particular RSA ciphertexts C. In this special case, The factorization of the RSA modulus N is not needed. In the next section, we shall consider the more general quantum attack by factoring N.

5.4 Quantum Integer Factorization Attack

Instead of finding the order of C in $\mathbb{Z}_N^*$, one can take this further to a more general case: find the order of an element x in $\mathbb{Z}_N^*$, denoted by order(x, N), where N is the RSA modulus. Once the order of an element x in $\mathbb{Z}_N^*$ is found, and it is even, it will have a good chance to factor N, of course in polynomial-time, by just calculating

$$\left\{\gcd(x^{r/2}+1, N), \quad \gcd(x^{r/2}-1, N)\right\}.$$

For instance, for $x = 3$, $r = 14718$ and $N = 179359$, we have

$$\left\{\gcd(3^{14718/2}+1, 179359), \quad \gcd(3^{14718/2}-1, 179359)\right\} = (67, 2677),$$

and hence the factorization of N:

$$N = 179359 = 67 \cdot 2677.$$

The following theorem shows that the probability for r to work is high.

Theorem 5.4.1. Let the odd integer $N > 1$ have exactly k distinct prime factors. For a randomly chosen $x \in \mathbb{Z}_N^*$ with multiplicative order r, the probability that r is even and that

$$x^{r/2} \not\equiv -1 \pmod{N}$$

is least $1 - 1/2^{k-1}$. More specifically, if N has just two prime factors (this is often the case for the RSA modulus N), then the probability is at least $1/2$.

Algorithm 5.4.1 (Quantum Algorithm for Integer Factorization). Given integers x and N, the algorithm will

– find the order of x, i.e., the smallest positive integer r such that

$$x^r \equiv 1 \pmod{N},$$

– find the prime factors of N and compute the decryption exponent d,
– decode the RSA message.

Assume the machine has two quantum registers: Register-1 and Register-2, which hold integers in binary form.

[1] (Initialization) Find a number q, a power of 2, say 2^t, with $N^2 < q < 2N^2$.

[2] (Preparation for quantum registers) Put in the first t-qubit register, Register-1, the uniform superposition of states representing numbers $a \pmod q$, and load Register-2 with all zeros. This leaves the machine in the state $|\Psi_1\rangle$:

$$|\Psi_1\rangle = \frac{1}{\sqrt{q}} \sum_{a=0}^{q-1} |a\rangle\,|0\rangle\,.$$

(Note that the joint state of both registers are represented by $|$ Register-1$\rangle$ and $|$ Register-2$\rangle$). What this step does is put each qubit in Register-1 into the superposition

$$\frac{1}{\sqrt{2}}(|0\rangle + |1\rangle)\,.$$

[3] (Base Selection) Choose a random $x \in [2, N-2]$ such that $\gcd(x, N) = 1$.

[4] (Power Creation) Fill in the second t-qubit register, Register-2, with powers $x^a \pmod N$. This leaves the machine in state $|\Psi_2\rangle$:

$$|\Psi_2\rangle = \frac{1}{\sqrt{q}} \sum_{a=0}^{q-1} |a\rangle\,|x^a \pmod N\rangle\,.$$

This step can be done reversibly since all the a's were kept in Register-1.

[5] (Perform a quantum FFT) Apply FFT on Register-1. The FFT maps each state $|a\rangle$ to

$$\frac{1}{\sqrt{q}} \sum_{c=0}^{q-1} \exp(2\pi i a c/q)\,|c\rangle\,.$$

That is, we apply the unitary matrix with the (a, c) entry equal to $\frac{1}{\sqrt{q}} \exp(2\pi i a c/q)$. This leaves the machine in the state $|\Psi_3\rangle$:

$$|\Psi_3\rangle = \frac{1}{q} \sum_{a=0}^{q-1} \sum_{c=0}^{q-1} \exp(2\pi i a c/q)\,|c\rangle\,|x^a \pmod N\rangle\,.$$

[6] (Periodicity Detection in x^a) Observe both $|c\rangle$ in Register-1 and $|x^a \pmod N\rangle$ in Register-2 of the machine, measure both arguments of this superposition, obtaining the values of $|c\rangle$ in the first argument and some $|x^k \pmod n\rangle$ as the answer for the second one ($0 < k < r$).

[7] (Extract r) Extract the required value of r. Given the pure state $|\Psi_3\rangle$, the probabilities of different results for this measurement will be given by the probability distribution:

$$\begin{aligned}
\text{Prob}(c, x^k \pmod N) &= \left| \frac{1}{q} \sum_{\substack{a=0 \\ x^a \equiv a^k \pmod N}}^{q-1} \exp(2\pi i ac/q) \right|^2 \\
&= \left| \frac{1}{q} \sum_{B=0}^{\lfloor (q-k-1)/r \rfloor} \exp(2\pi i (br+k)c/q) \right|^2 \\
&= \left| \frac{1}{q} \sum_{B=0}^{\lfloor (q-k-1)/r \rfloor} \exp(2\pi i b\{rc\}/q) \right|^2
\end{aligned}$$

where $\{rc\}$ is $rc \bmod N$. As showed in [287],

$$\begin{aligned}
\frac{-r}{2} \leq \{rc\} \leq \frac{-r}{2} \quad &\Longrightarrow \quad \frac{-r}{2} \leq rc - dq \leq \frac{-r}{2}, \text{ for some } d \\
&\Longrightarrow \quad \text{Prob}(c, x^k \pmod N) > \frac{1}{3r^2}.
\end{aligned}$$

then we have

$$\left| \frac{c}{q} - \frac{d}{r} \right| \leq \frac{1}{2q}.$$

Since $\frac{c}{q}$ were known, r can be obtained by the continued fraction expansion of $\frac{c}{q}$.

[8] (Resolution) If r is odd, go to Step [3] to start a new base. If r is even, then try to compute Once r is found, the factors of N can be *possibly*

$$\{\gcd(x^{r/2} - 1, N), \ \gcd(x^{r/2} + 1, N)\}$$

Hopefully, this will produce two factors of N.

[9] (Computing d) Once N is factored and p and q are found, then compute

$$d \equiv 1/e \pmod{(p-1)(q-1)}.$$

[10] (Code Break) As soon as d is found, and RSA ciphertext encrypted by the public-key (e, N), can be decrypted by this d as follows:

$$M \equiv C^d \pmod N.$$

Theorem 5.4.2 (Complexity of Quantum Factoring). Quantum factoring algorithm can factor the RSA modulus N and break the RSA system in time $\mathcal{O}((\log N)^{2+\epsilon})$.

Remark 5.4.1. The attack discussed in Algorithm 5.4.1 is more general than that in Algorithm 5.3.1. Algorithm 5.4.1 also implies that if a practical quantum computer can be built, then the RSA cryptosystem can be completely broken, and a quantum resistent cryptosystem must be developed and used to replace the RSA cryptosystem.

Example 5.4.1. On 19 December 2001, IBM made the first demonstration of the quantum factoring algorithm [317], that correctly identified 3 and 5 as the factors of 15. Although the answer may appear to be trivial, it may have good potential and practical implication. In this example, we show how to factor 15 quantum-mechanically [229].

[1] Choose at random $x = 7$ such that $\gcd(x, N) = 1$. We aim to find $r = \text{order}_{15} 7$ such that $7^r \equiv 1 \pmod{15}$.

[2] Initialize two four-qubit registers to state 0:

$$|\Psi_0\rangle = |0\rangle\,|0\rangle\,.$$

[3] Randomize the first register as follows:

$$|\Psi_0\rangle \to |\Psi_1\rangle = \frac{1}{\sqrt{2^t}} \sum_{k=0}^{2^t-1} |k\rangle\,|0\rangle\,.$$

[4] Unitarily compute the function $f(a) \equiv 13^a \pmod{15}$ as follows:

$$\begin{aligned}
|\Psi_1\rangle \to |\Psi_2\rangle &= \frac{1}{\sqrt{2^t}} \sum_{k=0}^{2^t-1} |k\rangle\,|13^k \pmod{15}\rangle \\
&= \frac{1}{\sqrt{2^t}} [\ |0\rangle\,|1\rangle + |1\rangle\,|7\rangle + |2\rangle\,|4\rangle + |3\rangle\,|13\rangle + \\
&\qquad |4\rangle\,|1\rangle + |5\rangle\,|7\rangle + |6\rangle\,|4\rangle + |7\rangle\,|13\rangle + \\
&\qquad |8\rangle\,|1\rangle + |9\rangle\,|7\rangle + |10\rangle\,|4\rangle + |11\rangle\,|13\rangle + \\
&\qquad + \cdots]
\end{aligned}$$

[5] We now apply the FFT to the second register and measure it (it can be done in the first), obtaining a random result from $1, 7, 4, 13$. Suppose we incidently get 4, then the state input to FFT would be

$$\sqrt{\frac{4}{2^t}}\ [\ |2\rangle + |6\rangle + |10\rangle + |14\rangle + \cdots]\,.$$

After applying FFT, some state

$$\sum_{\lambda} \alpha_\lambda\,|\lambda\rangle$$

with the probability distribution for $q = 2^t = 2048$ (see [229]). The final measurement gives $0, 512, 1024, 2048$, each with probability almost exactly $1/4$. Suppose $\lambda = 1536$ was obtained from the measurement. Then we compute the continued fraction expansion

$$\frac{\lambda}{q} = \frac{1536}{2048} = \frac{1}{1+\frac{1}{3}}, \quad \text{with convergents} \quad \left[0, 1, \frac{3}{4},\right]$$

Thus, $r = 4 = \text{order}_{15}(7)$. Therefore,

$$\gcd(x^{r/2} \pm 1, N) = \gcd(7^2 \pm 1, 15) = (5, 3).$$

Remark 5.4.2. Quantum factoring is still in its very earlier stage and will not threaten the security of RSA at least at present, as the current quantum computer can only factor a number with only 2 digits such as 15 which is essentially hopeless.

5.5 Quantum Discrete Logarithm Attack

Quantum algorithms can not only factor large integers in polynomial-time, but also take discrete logarithms in polynomial-time. In this section, we give a brief description of a quantum algorithm for solving the DLP in polynomial-time (provided that there is a practical quantum computer); with some modification, the algorithm can be used to solve the Elliptic Curve Discrete Logarithm (ECDLP).

Algorithm 5.5.1 (Quantum algorithm for discrete logarithms). Given $g, x \in \mathbb{N}$ and p prime. This algorithm will find the integer r such that $g^r \equiv x \pmod{p}$ if r exists. It uses three quantum registers.

[1] Find q a power of 2 such that q is close to p, that is, $p < q < 2p$.

[2] Put in the first two registers of the quantum computer the uniform superposition of all $|\,a\rangle$ and $|\,b\rangle \pmod{p-1}$, and compute $g^a x^{-b} \pmod{p}$ in the third register. This leaves the quantum computer in the state $|\,\Psi_1\rangle$:

$$|\,\Psi_1\rangle = \frac{1}{p-1}\sum_{a=0}^{p-2}\sum_{b=0}^{p-2}|\,a,\ b,\ g^a x^{-b} \pmod{p}\rangle\,. \tag{5.1}$$

[3] Use the Fourier transform A_q to map $|a\rangle \to |c\rangle$ and $|b\rangle \to |d\rangle$ with probability amplitude

$$\frac{1}{q}\exp\left(\frac{2\pi i}{q}(ac+bd)\right).$$

Thus, the state $|a,b\rangle$ will be changed to the state:

$$\frac{1}{q}\sum_{c=0}^{q-1}\sum_{d=0}^{q-1}\exp\left(\frac{2\pi i}{q}(ac+bd)\right)|c,\ d\rangle. \tag{5.2}$$

This leaves the machine in the state $|\Psi_2\rangle$:

$$|\Psi_2\rangle = \frac{1}{(p-1)q}\sum_{a,b=0}^{p-2}\sum_{c,d=0}^{q-1}\exp\left(\frac{2\pi i}{q}(ac+bd)\right)\left|c,\ d,\ g^a x^{-b} \pmod p\right\rangle. \tag{5.3}$$

[4] Observe the state of the quantum computer and extract the required information. The probability of observing a state $\left|c,\ d,\ g^k \pmod p\right\rangle$ is

$$\left|\frac{1}{(p-1)q}\sum_{a,b}\exp\left(\frac{2\pi i}{q}(ac+bd)\right)\right|^2 \tag{5.4}$$

where the sum is over all (a,b) such that

$$a - rb \equiv k \pmod{p-1}. \tag{5.5}$$

The better outputs (observed states) we get, the more chance of deducing r we will have; readers are referred to [288] for a justification.

Remark 5.5.1. The quantum discrete logarithm algorithm is an extension of the quantum factoring algorithms. It is noted that the the quantum discrete logarithm algorithm can also be extended to solve the Elliptic Curve Discrete Logarithm (ECDLP). Thus, if a practical quantum computer becomes available, then essentially all the currently used public-key cryptosystems based on IFP, DLP and ECDLP will not be secure. However, as we mentioned earlier, a practical quantum computer may never be able to built, at least at present or in the next 30 years. Thus RSA should be secure for at least 30 years.

5.6 Chapter Notes and Further Reading

Quantum algorithms seem very well suited for attacking all those cryptosystems whose security is based on the integer factorization problem (IFP), the discrete logarithm problem (DLP), and the elliptic curve discrete logarithm problem (ECDLP). Unfortunately, the security of most of the widely used cryptosystems are based, one way or another, on these three types of intractable number-theoretic problems. So if a practical quantum computer can be built, all the IFP-based, DLP-based, and ECDLP-based cryptosystems will become useless, and as a consequence, we have to replace them with some quantum resistance cryptosystems. Fortunately (from a cryptographic point of view but unfortunately from a computer science point of view), practical computers may never be built or at least may need a very long time before they can be developed. Thus, all the IFP, DLP, and ECDLP based cryptosystems will remain to be useful for some time. Particularly, for RSA, it may still be useful for another 30 years, making its possible total life for about 60 years.

There are many good comprehensive references on quantum computation (particularly on quantum factoring which we are interested in) such as [130], [200], [229], [239], and [333]. Shor's original paper on quantum algorithms for factoring and discrete logarithms may be found in ([288]. The experimental realization of Shor's quantum factoring algorithm using nuclear magnetic resonance at IBM Almaden Research Center in California for factoring $15 = 3 \cdot 5$ is reported in a *Nature* paper [317]. Other interesting papers on quantum computing/factoring in the most prestigious journal *Science* and *Nature* include [28], [119], and [68]. Feynman's original idea about quantum computers can be found in [111] and [112]. The quantum algorithm for the elliptic curve discrete logarithm problem may be found in [250]. An elementary algorithmic description of quantum factoring/computing can be found in the textbook [92] whereas a more number-theoretic oriented description of quantum factoring/computing can be found in [335].

6. Simple Elementary Attacks

There is always more spirit in attack than in defence.

TITUS LIVIUS (59 BC–17 AD)
Roman Author and Historian

6.1 Introduction

The previous three chapters have been mainly concerned with the *direct algorithmic attacks*, both classical or quantum-mechanical, on the RSA modulus N, namely integer factorization attacks, discrete logarithm attacks, and quantum mechanical attacks. It is interesting to note that most successful attacks on RSA, however, are not based on factoring the modulus N and do not result from the use of insufficiently large N. Rather, they exploit the mathematical weakness of the RSA algorithm or the improper use of the RSA system, such as common modulus, and lower exponents, etc. These attacks are called the *indirect algorithmic attacks.* Some non-algorithmic attacks on RSA also exist; they exploit specific *hardware implementation issues* of the RSA algorithm/system; nothing to do with the inherent weakness of the algorithm/system itself. For example, by carefully measuring the power consumption or timing of a cryptographic device (e.g., a smartcard or a computer) it takes on performing a particular cryptographic algorithm, it is possible to extract the secret-key of the cryptographic device. These attacks are called *side-channel attacks.* In the rest of the book, we shall concentrate on the indirect algorithmic attacks and the side-channel attacks.

The strategies of attacks and anti-attacks (defenses) of an encryption is the same as that of missiles and anti-missiles of the military weapons. From a cryptanalytic point of view, it would be nice to find some unaware weak points of a cryptographic system so that the system can be broken. From a cryptographic point of view, however, it would be nice to design a convenient, efficient and unconditionally unbreakable cryptographic system. Unfortunately, it is very difficult to satisfy both. On the one hand, unconditionally unbreakable cryptographic systems, such as the One-Time Pads (OTPs) are indeed possible, but are largely impractical and hence not so useful, due to the cost, inefficiency and the too frequently changing of the keys. On the other hand, the conditional unbreakable cryptographic systems, such as RSA, are practical, efficient, convenient, but it is a risk to use such cryptosystems, since their security is based on some unproven conjectures such as the RSA conjecture, and since there may exist some unaware weaknesses and unaware attacks on the systems that may have been kept a secret and may have never been reported and disclosed to the public-it is well possible! In this chapter we shall study some simple but indirect elementary (and of course known) mathematical/algorithmic attacks on the RSA system.

6.2 Guessing Plaintext Attacks

Our first type of simple elementary algorithmic attacks is concerned with the attacks on guessing the values of the plaintexts M. Suppose (e, N, C) is given, and the cryptanalyst, Eve, wishes to find M. That is,

$$\{e, N, C \equiv M^e \pmod{N}\} \xrightarrow[\text{guessing } M]{\text{find}} \{M\}.$$

If the plaintext space $\mathcal{M} = \{M_1, M_2, \cdots, M_k\}$ is small or predicable, Eve can decrypt C by simply encrypting all possible plaintext messages $M_1, M_2, \cdots, M_k$ to get $C_1', C_2', \cdots, C_k'$, and check, at each step, if $C_i' = C$. If yes, then $M = M_i$, the plaintext M is found. This simple process can be described as follows:

$$\begin{array}{ll}
C_1' \equiv (M_1)^e \pmod{N}, & \text{if } C_1' = C, \text{ then } M = M_1, \\
C_2' \equiv (M_2)^e \pmod{N}, & \text{if } C_2' = C, \text{ then } M = M_2, \\
\vdots & \vdots \\
C_k' \equiv (M_k)^e \pmod{N}, & \text{if } C_1' = C, \text{ then } M = M_k.
\end{array}$$

This attack is known as *forward search attack*, or *guessing plaintext attack*. The attack will be impractical if the message space $\mathcal{M}$ is large. So to prevent such an attack, the message space $\mathcal{M}$ is necessarily very large.

A closely related attack to the forward search attack is the *short plaintext attack*. If the plaintext message M is small although the corresponding C can be as big as N (this is the general case for public-key cryptography as it is usually only used to encrypt short massages particularly the encryption keys used for secret-key cryptographic systems such as DES and AES [315]), then the cryptanalyst can perform two sequences of the operations as follows:

$$\begin{aligned} U &\equiv Cx^{-e} \pmod N, \quad \text{for all } 1 \le x \le 19^9 \\ V &\equiv y^e \pmod N, \quad \text{for all } 1 \le y \le 19^9 \end{aligned}$$

If for some of the pair (x, y), we have $U = V$, then $C \equiv (xy)^e \pmod N$. Thus $M = xy$. This attack is much more efficient than the forward attack that would try all 10^{17} possible values of M, because it only needs to perform 2×10^9 computations and to compare the elements in the two sequences up to 10^9 times.

The above two attacks can be easily prevented by a *salting process* (i.e., appending some random digits to the plaintext message M prior to encryption), or by a *padding process* (i.e., adjoin some random digits to the beginning and the end of the plaintext message M prior to encryption) such as the one discussed in [22], so that a large random plaintext M can be formed; these randomly added digits can be simply removed after decryption.

6.3 Blinding Attack on RSA Signatures

Our second type of simple elementary algorithmic attacks is on RSA signatures. Suppose (e, N, M) is given, and the cryptanalyst, Eve, wishes to find the digital signature S. That is,

$$\{e, N, M \equiv S^e \pmod N\} \xrightarrow[\text{forging } S]{\text{find}} \{S\}. \tag{6.1}$$

As the RSA function enjoys certain kind of *self-reducibility* [264], which, on the one hand is good (as it provides assurance that all random ciphertexts are equally hard to decrypt) but on the other hand is bad (as it provides an avenue for an attacker, Eve, to gain information about the decryption of one ciphertext from the decryption of other ciphertexts). The following attack, called the *blinding attack* [37] is based, unfortunately, on this self-reducibility and can be used to obtain someone's valid digital signature.

Let (e, N) and (d, N) be Bob's public and secret keys, respectively. Suppose the cryptanalyst, Eve, wants to know Bob's signature S on a message $M \in \mathbb{Z}_N^*$, which is computed by:

$$S \equiv M^d \pmod N.$$

Then Eve can try the following:

[1] Eve picks up a random number $r \in \mathbb{Z}_N^*$, and computes $M' \equiv r^e M \pmod{N}$.

[2] Even asks Bob to sign the random message (looks like a hashed value, as it usually should be) M'.

[3] Suppose Bob is willing to sign the message M', which means that Eve can get

$$S' \equiv (M')^d \pmod{N}.$$

[4] Now it is plain for Eve to get Bob's valid signature S as follows:

$$S \equiv S'/r \pmod{N},$$

which is so because

$$\begin{aligned} S^e &\equiv (S'/r)^e \\ &\equiv (S')^e/r^e \\ &\equiv ((M')^d)^e/r^e \\ &\equiv M'/r^e \\ &\equiv (r^e M)/r^e \\ &\equiv M \pmod{N}. \end{aligned}$$

Thus, Eve can forge Bob's valid signature without knowing his private exponent d, and Bob will not detect the forgery since $M \equiv S^e \pmod{N}$, as we have just showed.

Again, this *chosen plaintext attack* can be avoided by using random padding techniques. Note that the random padding techniques are also countermeasures against the following *chosen-ciphertext attack* [269]. Suppose the cryptanalyst, Eve, intercepts a ciphertext C from Bob to Alice. Then Eve chooses at random a positive integer r, computes $\tilde{M} \equiv C \cdot r^e \pmod{N}$ and sends it to Alice. Alice then decrypts the ciphertext $\tilde{C} \equiv \tilde{M}^e \equiv C^d \cdot r \pmod{N}$. Suppose now Eve can get this $\tilde{C}$, then she can get the original plaintext M by computing

$$M \equiv r^{-1} C^d \cdot r \pmod{N}.$$

6.4 Guessing $\phi(N)$ Attack

If one can guess the value of $\phi(N)$, one can recover the RSA plaintext M from its corresponding ciphertext C in polynomial-time. That is,

$$\phi(N) \stackrel{\mathcal{P}}{\Longrightarrow} \{M\}. \tag{6.2}$$

First of all, we show that the computation of $\phi(N)$ and the factorization of N, IFP(N), are deterministic polynomial-time equivalent.

Theorem 6.4.1. (The equivalence of $\phi(N)$ and IFP(N))

$$\phi(N) \stackrel{\mathcal{P}}{\Longleftrightarrow} \text{IFP}(N). \tag{6.3}$$

Proof. Note first that if $(N, \phi(N))$ is known and N is assumed to be the product of two primes p and q, then N can be easily factored. Assume

$$N = pq,$$

then

$$\phi(N) = (p-1)(q-1),$$

thus

$$pq - p - q + 1 - \phi(N) = 0 \tag{6.4}$$

substituting $q = n/p$ into (6.4) gives

$$p^2 - (N - \phi(N) + 1)p + N = 0. \tag{6.5}$$

Let $A = N - \phi(N) + 1$, then

$$(p, q) = \frac{A \pm \sqrt{A^2 - 4N}}{2}$$

will be the two roots of (6.5), and hence, the two prime factors of N.

On the other hand, if the two prime factors p and q of N are known, then $\phi(N) = (p-1)(q-1)$ immediately from

$$\phi(N) = N \prod_{i=1}^{k} p_i^{\alpha_i}$$

if

$$N = \prod_{i=1}^{k} p_i.$$

□

What this theorem says is that if an enemy cryptanalyst could compute $\phi(N)$ then he could break RSA by computing d as the multiplicative inverse of e modulo $\phi(N)$. That is, $d \equiv 1/e \pmod{\phi(N)}$. On the other hand, the knowledge of $\phi(N)$ can lead to an easy way of factoring N, since

$$
\begin{aligned}
p+q &= n-\phi(N)+1, \\
(p-q)^2 &= (p+q)^2-4n, \\
p &= \frac{(p+q)+(p-q)}{2}, \\
q &= \frac{(p+q)-(p-q)}{2}.
\end{aligned}
$$

In other words, computing $\phi(N)$ is no easier than factoring N.

Example 6.4.1. Let

$$N = 74153950911911911.$$

Suppose the cryptanalyst knows by guessing, interception or whatever that

$$\phi(N) = 74153950339832712.$$

Then

$$
\begin{aligned}
A &= N-\phi(N)+1 \\
&= 74153950911911911 - 74153950339832712 + 1 \\
&= 572079200
\end{aligned}
$$

Thus

$$p^2 - 572079200p + 74153950911911911 = 0.$$

Solving this equation gives the two roots

$$\{p, q\} = \{198491317,\ 373587883\},$$

and hence the complete prime factorization of N

$$
\begin{aligned}
N &= 74153950911911911 \\
&= 198491317 \cdot 373587883.
\end{aligned}
$$

Theorem 6.4.2. The RSA encryption is breakable in polynomial-time if the cryptanalyst knows $\phi(N)$. That is,

$$\phi(N) \overset{\mathcal{P}}{\Longrightarrow} \text{RSA}(M). \tag{6.6}$$

Proof. If $\phi(N)$ is known, then $d \equiv 1/e \pmod{\phi(N)}$, hence recovers M from C: $M \equiv C^d \pmod{N}$. □

Thus, we have:

$$\text{IFP}(N) \overset{\mathcal{P}}{\Longleftrightarrow} \phi(N) \overset{\mathcal{P}}{\Longrightarrow} \text{RSA}(M). \tag{6.7}$$

Thus, breaking the RSA encryption by computing $\phi(N)$ is no easier than breaking the RSA encryption by factoring N. However, if someone can intelligently and efficiently guess/find the value of $\phi(N)$, or someone has already known $\phi(N)$ by some means, then he can break RSA completely without factoring.

6.5 Guessing d Attack

If N is large and d is chosen from a large set, then a cryptanalyst should not be able to determine d any easier than he can factor N. We first observe that factoring N is *deterministic polynomial-time equivalent* to computing the RSA private exponent d. That is,

$$\text{IFP}(N) \overset{\mathcal{P}}{\Longleftrightarrow} \{d\},$$

or

$$\text{IFP}(N) \overset{\mathcal{P}}{\Longleftrightarrow} \text{RSA}(d).$$

Theorem 6.5.1. If the prime factorization of the RSA modulus N is known, then d can be calculated in polynomial-time. That is,

$$\text{IFP}(N) \overset{\mathcal{P}}{\Longrightarrow} \{d\}. \tag{6.8}$$

Proof. If $N = pq$ is known, then

$$d \equiv 1/e \pmod{(p-1)(q-1)}$$

since e is also known. As the modular inverse $1/e$ can be done in polynomial-time by the extended Euclid's algorithm, d can be calculated in polynomial-time.

□

Now we show that given the RSA private exponent d, the prime factorization of N can be done in polynomial time.

Theorem 6.5.2 (Coron and May [85]). Let $N = pq$ with p and q prime numbers. Let also e and d be the public and private exponent, respectively, satisfying $ed \equiv 1 \pmod{\phi(N)}$.

(1) If p and q are with the same bit size and $1 < ed \leq N^{3/2}$, then given (N, e, d), the prime factorization of N can be computed deterministically in time $\mathcal{O}((\log N)^2)$.

(2) If p and q are with the same bit size and $1 < ed \leq N^2$, then given (N, e, d), the prime factorization of N can be computed deterministically in time $\mathcal{O}((\log N)^9)$.

(3) Let β and $0 < \delta \leq 1/2$ be real numbers such that $2\beta\delta(1-\delta) \leq 1$. Let $N = pq$ with p and q primes such that $p < N^\delta$ and $q < 2N^{1-\delta}$. Let $1 < ed \leq N^\beta$. Then given (N, e, d), the prime factorization of N can be computed deterministically in time $\mathcal{O}((\log N)^9)$.

Proof. The results follow by applying Coppersmith's technique [78] of finding small solutions to the univariable modular polynomial equations using lattice reduction [188]. For more details, see [85]. □

Corollary 6.5.1. If d is known, then the prime factorization N can be found in deterministic polynomial-time. That is,

$$\{d\} \overset{\mathcal{P}}{\Longrightarrow} \mathrm{IFP}(N). \tag{6.9}$$

Combining Theorem 6.5.1 and Corollary 6.5.1, we have

Theorem 6.5.3 (The equivalence of RSA(d) and IFP(N)).
Computing the private exponent d by giving the prime factorization N and computing the prime factorization of N by giving the private exponent d are deterministic polynomial-time equivalent. That is,

$$\{d\} \overset{\mathcal{P}}{\Longleftrightarrow} \mathrm{IFP}(N). \tag{6.10}$$

Remark 6.5.1. It was known, as soon as the RSA cryptographic system was designed in 1977, that if the RSA private exponent d is given, then the prime factorization of the RSA modulus N can be computed in *random* polynomial-time by using Miller's techniques developned in 1976 [212]. That is,

$$\{d\} \overset{\mathcal{RP}}{\Longrightarrow} \mathrm{IFP}(N). \tag{6.11}$$

The proof is proceed as follows. First note that

$$ed \equiv 1 \pmod{\phi(N)}.$$

then

$$ed = t\phi(N) + 1, t \in \mathbb{Z}.$$

Pick up at random $x \in \mathbb{Z}_{\neq 0}$, this is guaranteed to satisfy

$$x^{ed-1} \equiv 1 \pmod N.$$

Then computing y_1 of 1 modulo N yields:

$$y_1 \equiv \sqrt{x^{ed-1}} \equiv x^{(ed-1)/2} \pmod N.$$

Therefore,

$$y_1^2 - 1 \equiv 0 \pmod{N}.$$

Thus, N can be factorized by computing

$$\gcd(y_1 \pm 1, N).$$

But this will only work when $y_1 \not\equiv \pm 1 \pmod{N}$. Suppose we are unlucky and obtain $y_1 \equiv \pm 1 \pmod{N}$ rather than a factor of N. If $y_1 \equiv -1 \pmod{N}$, we return to the beginning and pick up another integer x. If $y_1 \equiv 1 \pmod{N}$, we take another square root of one via

$$\begin{aligned} y_2 &\equiv \sqrt{y_1} \\ &\equiv x^{(ed-1)/4} \pmod{N}. \end{aligned}$$

Hence,

$$y_2^2 - 1 \equiv 0 \pmod{N}.$$

Computing

$$\gcd(y_2 \pm 1, N).$$

Again, this will give a factor of N unless

$$y_2 \equiv \pm 1 \pmod{N}.$$

If we are unlucky, repeat the above process again (and again) until we have either factorized N or found $2 \nmid (ed-1)/2^s$ for some $s \in \mathbb{Z}$. Clearly, the above process can be done in random polynomial-time. On the other hand, if IFP(N) is known then one can easily and deterministically find d in polynomial-time just by computing $d \equiv 1/e \pmod{(p-1)(q-1)}$.

Finally, we show that if the prime factorization of the RSA modulus N is known, the RSA plaintext M can be computed in polynomial-time from the corresponding ciphertext C.

Theorem 6.5.4. If the prime factorization of the RSA modulus N is known, then RSA can be broken in polynomial-time. That is,

$$\text{IFP}(N) \stackrel{\mathcal{P}}{\Longrightarrow} M. \tag{6.12}$$

Proof. By Theorem 6.5.1, if the prime factorization of N is known, then d can be calculated in polynomial-time. Once d is found, then the following decryption process

$$M \equiv C^d \pmod{N},$$

can be done in polynomial-time, as (C, N) is known. □

This is the same as to say that computing d from the public-key (e, N) is as hard as factoring the modulus N.

Example 6.5.1. Let

$$\begin{aligned} e &= 17579, \\ N &= 6397848687952714385883141504 1, \\ d &= 10663687727232084624328285019. \end{aligned}$$

We wish to find the prime factors p and q of N from the secret-key d. We follow the procedure given in the proof of Theorem 6.5.1. Let $x = 2$ and perform the following computations:

```
for i from 2 to 100 do
    s_i := (e * d - 1)/i
    y_i := 2^t_i mod N
    print(i, s_i, y_i, gcd(y_i - 1, N))
    end_do
```

We find that only those numbers when $i = 13, 26, 39, 52, 65, 78, 81, 91$ for $i \leq 100$ are lucky and give rise to $y_i \not\equiv 1 \pmod{N}$ and hence each leads to the complete prime factorization of $N = 145295143558111 \cdot 440334654777631$:

$$\begin{cases} s_{13} = (ed-1)/13 = 14419766658231755047005147873000 \\ y_{13} \equiv x^{s_{13}} \pmod{N} = 8844029226054068856172959205 \\ \gcd(y_{13}-1, N) = 440334654777631 \end{cases}$$

$$\begin{cases} s_{26} = (ed-1)/26 = 7209883329115877523502573936500 \\ y_{26} \equiv x^{s_{26}} \pmod{N} = 27598022604590536915466802 86 \\ \gcd(y_{13}-1, N) = 440334654777631 \end{cases}$$

$$\begin{cases} s_{39} = (ed-1)/39 = 4806588886077251682335049291000 \\ y_{39} \equiv x^{s_{39}} \pmod{N} = 14087419621751444280492087156 \\ \gcd(y_{39}-1, N) = 440334654777631 \end{cases}$$

$$\begin{cases} s_{52} = (ed-1)/52 = 3604941664557938761751286968250 \\ y_{52} \equiv x^{s_{52}} \pmod{N} = 33661945935813861391560228598 \\ \gcd(y_{52}-1, N) = 440334654777631 \end{cases}$$

$$\begin{cases} s_{65} = (ed-1)/65 = 2883953331646351009401029574600 \\ y_{65} \equiv x^{s_{65}} \pmod{N} = 59017354193359494219779573422 \\ \gcd(y_{65}-1, N) = 440334654777631 \end{cases}$$

$$\begin{cases} s_{78} = (ed-1)/78 = 2403294443038625841167524645500 \\ y_{78} \equiv x^{s_{78}} \pmod{N} = 9564171122158182570859195332 \\ \gcd(y_{78}-1, N) = 440334654777631 \end{cases}$$

$$\begin{cases} s_{81} = (ed-1)/81 = 23142835377408989581613200290000 \\ y_{81} \equiv x^{s_{81}} \pmod{N} = 355905005236966211763919095559 \\ \gcd(y_{81}-1, N) = 145295143558111 \end{cases}$$

$$\begin{cases} s_{91} = (ed-1)/91 = 20599666654616792924293068390000 \\ y_{91} \equiv x^{s_{91}} \pmod{N} = 425913211637207795520959446366 \\ \gcd(y_{91}-1, N) = 440334654777631 \end{cases}$$

The unlucky values of i in the range give rise to either $y_i \equiv 1 \pmod{N}$ or $s_i \notin \mathbb{Z}$.

Thus, to avoid the guessing d attack, N must be large, and d should be chosen from a large set such that the cryptanalyst cannot easily choose the correct d from the large set.

6.6 e^{th} Root Attack

The RSA encryption works as follows:

$$C \equiv M^e \pmod{N}. \tag{6.13}$$

Thus, the most straightforward attack on the RSA encryption would be taking the eth root of the ciphertext C modulo N:

$$M \equiv \sqrt[e]{C} \pmod{N}. \tag{6.14}$$

So, to break the RSA encryption it suffices to solve the Root Find Problem (RFP) defined in (6.14). Clearly, if the eth root of C is given, M can be computed in polynomial-time. That is,

$$\text{RFP}(C) \stackrel{\mathcal{P}}{\Longrightarrow} \{M\}, \tag{6.15}$$

or

$$\text{RFP}(C) \stackrel{\mathcal{P}}{\Longrightarrow} \text{RSA}(M). \tag{6.16}$$

It is evident that in (6.13), once $\langle N, e, C \rangle$ is given, we could try to substitute the values for $M = 0, 1, 2, \cdots$ until a correct M is found. In theory, it is possible to enumerate all elements of $\mathbb{Z}_N^*$, since $\mathbb{Z}_N^*$ is a finite set, but in practice, it is impossible when N is large. However, if $\phi(N)$ is known, then we can compute the eth root of C modulo N fairly easily, in polynomial-time.

Theorem 6.6.1.

$$\phi(N) \stackrel{\mathcal{P}}{\Longrightarrow} \text{RFP}(C). \tag{6.17}$$

Proof. Note that in RSA,

$$C \equiv M^e \pmod{N}, \; M \equiv M^d \pmod{N}, \; ed \equiv 1 \pmod{\phi(N)}.$$

If $\phi(N)$ is known, then the positive integers d and k in the equation

$$ed - k\phi(N) = 1, \; k \in \mathbb{Z}$$

can be determined by the extended Euclid's algorithm which runs in polynomial-time. Now use the value of d to get

$$M \equiv C^d \pmod{N}$$

which is then the required value for $\sqrt[e]{C} \bmod N$, since

$$\begin{aligned} M^e &\equiv (C^d)^e \\ &\equiv C^{ed} \\ &\equiv C^{1+k\phi(N)} \\ &\equiv C \cdot C^{\phi(N)} \\ &\equiv C \pmod{N}. \end{aligned}$$

Example 6.6.1. Let $C \equiv M^{131} \pmod{1073}$, find M such that

$$M \equiv \sqrt[131]{758} \pmod{1073}.$$

or equivalently,

$$M^{131} \equiv 758 \pmod{1073}.$$

Assume $\phi(1073) = 1008$ is known. Then we can solve the linear Diophantine equation in d and k:

$$131d - 1008k = 1.$$

Since $131/1008$ can be expanded as a finite continued fraction

$$\begin{aligned} \frac{131}{1008} &= 0 + \cfrac{1}{7 + \cfrac{1}{2 + \cfrac{1}{3 + \cfrac{1}{1 + \cfrac{1}{1 + \cfrac{1}{1 + \cfrac{1}{3}}}}}}} \\ &= [0, 7, 2, 3, 1, 1, 1, 1, 3] \end{aligned}$$

with convergents:

$$\left[0, \frac{1}{7}, \frac{1}{8}, \frac{3}{23}, \frac{10}{77}, \frac{13}{100}, \frac{23}{177}, \frac{36}{277}, \frac{131}{1008}\right]$$

then

$$\begin{aligned} d &= (-1)^{n-1} q_{n-1} = (-1)^{8-1} \cdot 277 = -277 \\ k &= (-1)^{n-1} p_{n-1} = (-1)^{8-1} \cdot 36 = -36, \end{aligned}$$

satisfying

$$131(-277) - 1008(-36) = 1.$$

Now compute

$$\begin{aligned} M &\equiv (M^{131})^d \\ &\equiv 758^{-277} \\ &\equiv 1/758^{277} \\ &\equiv 1/875 \\ &\equiv 905 \pmod{1073}. \end{aligned}$$

However, to find $\phi(N)$, the only efficient way we know of is factoring N. Thus, breaking the RSA encryption by computing $\sqrt[e]{C} \bmod N$, assuming $\phi(N)$ is known, is no easier than breaking the encryption by factoring N.

Theorem 6.6.2.

$$\text{IFP}(N) \overset{\mathcal{P}}{\Longleftrightarrow} \phi(N) \overset{\mathcal{P}}{\Longrightarrow} \text{RFP}(C) \overset{\mathcal{P}}{\Longrightarrow} M. \tag{6.18}$$

As we mentioned earlier, it is conjectured that *any method of breaking the RSA cryptosystem must be as difficult as factoring*, but this has never been proved or disproved, it may well be possible to break RSA without factoring, particularly when the public exponent e is small.

6.7 Common Modulus Attack

The four RSA parameters $\{d, p, q, \phi(N)\}$ form the RSA trap-door. These four pieces of information are equally important. Knowledge of any one of them reveals the knowledge of the remaining three, and hence break the RSA encryption completely. If RSA is not used properly, however, it may well be possible to break the RSA encryption without use of any knowledge of $\{d, p, q, \phi(N)\}$. One such improper use is the use of common modulus N

in RSA encryption. Suppose that Bob sends Alice two ciphertexts C_1 and C_2 as follows:

$$\begin{aligned} C_1 &\equiv M^{e_1} \pmod{N} \\ C_2 &\equiv M^{e_2} \pmod{N} \end{aligned}$$

where $\gcd(e_1, e_2) = 1$. Then as the following theorem shows, Eve can recover the plaintext M without factoring N or without using any of the trap-door information $\{d, p, q, \phi(N)\}$.

Theorem 6.7.1. Let $N_1 = N_2$ and $M_1 = M_2$ but $e_1 \neq e_2$ and $\gcd(e_1, e_2) = 1$ such that

$$\begin{aligned} C_1 &\equiv M^{e_1} \pmod{N} \\ C_2 &\equiv M^{e_2} \pmod{N} \end{aligned}$$

Then M can be recovered easily; that is,

$$\{[C_1, e_1, N],\ [C_2, e_2, N]\} \overset{\mathcal{P}}{\Longrightarrow} \{M\} \tag{6.19}$$

Proof. Since $\gcd(e_1, e_2) = 1$, then $e_1 x + e_2 y = 1$ with $x, y \in \mathbb{Z}$, which can be done by the extended Euclid's algorithm (or the equivalent continued fraction algorithm) in polynomial-time. Thus,

$$\begin{aligned} C_1^x C_2^y &\equiv (M_1^{e_1})^x (M_2^{e_2})^y \\ &\equiv M^{e_1 x + e_2 y} \\ &\equiv M \pmod{N} \end{aligned}$$

□

Example 6.7.1. Let

$$\begin{aligned} e_1 &= 9007, \\ e_2 &= 65537, \\ M &= 19050321180920251905182209030519. \\ N &= 114381625757888867669235779976146612010218296721242362\_ \\ &\quad 562561842935706935245733897830597123563958705058989075\_ \\ &\quad 147599290026879543541 \end{aligned}$$

Then

$$
\begin{aligned}
C_1 &\equiv M^{e_1} \bmod N \\
&\equiv 1042022509411962384136383826079741257744490847249295 9\_ \\
&\quad 1257433745889265297771717182413024642938078351979089 9\_ \\
&\quad 453434074641613779772 12 \\
C_2 &\equiv M^{e_2} \bmod N \\
&\equiv 7645275072918870018071997051754457471094475731790989 6\_ \\
&\quad 0413409874882855731902807834803090849780215633964907 5\_ \\
&\quad 975060051949607130434 8
\end{aligned}
$$

Now we determine x and y in

$$9007x + 65537y = 1.$$

First, we get the continued fraction expansion of $9007/65537$ as follows:

$$9007/65537 = [0, 7, 3, 1, 1, 1, 1, 1, 2, 1, 1, 1, 1, 2, 7].$$

Then we get the convergents of the continued fraction of $9007/65537$ as follows:

$$\left[0, \frac{1}{7}, \frac{3}{22}, \frac{4}{29}, \frac{7}{51}, \frac{11}{80}, \frac{18}{131}, \frac{29}{211}, \frac{76}{553}, \frac{105}{764}, \frac{181}{1317}, \frac{286}{2081}, \frac{467}{3398}, \frac{1220}{8877}, \frac{9007}{65537}\right]$$

Thus,

$$\left\{ \begin{array}{l} x = (-1)^{n-1}q_{n-1} = (-1)^{13}8877 = -8877, \\ y = (-1)^n p_{n-1} = (-1)^{14}1220 = 1220. \end{array} \right.$$

Therefore,

$$
\begin{aligned}
M &\equiv C_1^x C_2^y \\
&\equiv 1042022509411962384136383826079741257744490847249295 9\_ \\
&\quad 1257433745889265297771717182413024642938078351979089 9\_ \\
&\quad 45343407464161377977212^{-8877} \cdot \\
&\quad 7645275072918870018071997051754457471094475731790989 6\_ \\
&\quad 0413409874882855731902807834803090849780215633964907 5\_ \\
&\quad 9750600519496071304348^{1220} \\
&\equiv 19050321180920251905182209030519 \pmod N
\end{aligned}
$$

So, we can recover the plaintext M without factoring N and/or using any of the trap-door information $d, p, q, \phi(N)$.

This attack suggests that to defend RSA, one should never use common modulus in RSA encryption.

6.8 Fixed-Point Attack

In this section we introduce one more simple elementary attack on RSA, the fixed-point attack, which does not rely on the factorization of N and the use any trap-door information. The fixed-point attack is also called cyclic attack or *superencryption attack*. This attack was discovered by Simmons and Norris in 1977, as soon as the RSA cryptographic system was invented.

Definition 6.8.1. Let $0 \leq x < N$. If

$$x^{e^k} \equiv x \pmod{N}, \; k \in \mathbb{Z}^+ \tag{6.20}$$

then x is called the fixed-point of RSA(e, N) and k is the order of the fixed-point.

Theorem 6.8.1. Let C to be the fixed-point of RSA(e, N) with order k:

$$C^{e^k} \equiv C \pmod{N}, \; k \in \mathbb{Z}^+$$

then

$$C^{e^{k-1}} \equiv M \pmod{N}, \; k \in \mathbb{Z}^+$$

Proof. Since the RSA encryption $C \equiv M^e \pmod{N}$ is a permutation on the message space $\{0, 1, 2, \cdots, N-1\}$, such an integer (fixed-point) $C^{e^k} \equiv C \pmod{N}$ must exist. For the same reason, it must be the case that

$$C^{e^{k-1}} \equiv M \pmod{N},$$

since

$$\begin{aligned}
& C^{e^k} \equiv C \pmod{N} \\
\Longrightarrow \quad & C^{e^k} \equiv M^e \pmod{N} \\
\Longrightarrow \quad & C^{e^{k-1}e} \equiv M^e \pmod{N} \\
\Longrightarrow \quad & (C^{e^{k-1}})^e \equiv M^e \pmod{N} \\
\Longrightarrow \quad & C^{e^{k-1}} \equiv M \pmod{N}.
\end{aligned}$$

□

Theorem 6.8.1 provides a straightforward attack on RSA just by computing the sequence of numbers, modulo N:

$$\begin{array}{cccccc}
C^e & C^{e^2} & C^{e^3} & \cdots & C^{e^{k-1}} & C^{e^k} \\
 & & & & \Uparrow & \Downarrow \\
 & & & & M & C
\end{array}$$

That is, as soon as $C^{e^k} \bmod N = C$ is obtained, we stop the computation, and pick the last second number $C^{e^{k-1}} \bmod N$, which must be the RSA(M) (Depicted in Figure 6.1).

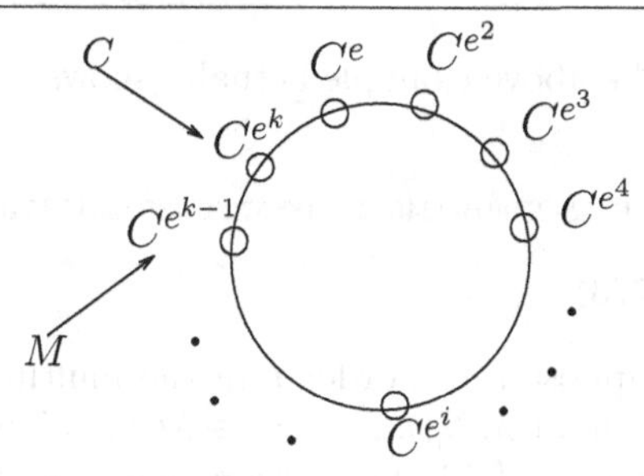

Figure 6.1. Fixed-Point Attack

Example 6.8.1. We illustrate the fixed-point attack by selecting an encryption example from the original RSA paper (Page 124 of [262]), where

$$\begin{aligned} e &= 17 \\ N &= 2773 \\ C &= 2342. \end{aligned}$$

We construct the sequence $C^{e^k} \bmod N$ for $k = 1, 2, 3, \cdots$ using the following process

```
e ← 17
N ← 2773
C ← 2342
for k from 1 to 100 do
    A ← C^(e^k) mod 2773
    if A = C then M ← C^(e^(k-1)) mod 2773
    print(k,M,C);
```

as follows:

2365	1157	2018	985	1421	2101	1664	2047	1539	980
1310	1103	1893	1629	2608	218	1185	1039	602	513
772	744	720	2755	890	2160	2549	926	536	449
2667	2578	182	2278	248	454	1480	1393	2313	2637
2247	1688	1900	2342						
		⇑	⇓						
		M	C						

Since $C \equiv 2342^{17^{44}} \pmod{2773}$ is the fixed-point of RSA(17, 2773) with $k = 44$, then $M \equiv C^{17^{43}} \equiv 1900 \pmod{2773}$ is the plaintext of $C = 2342$. Indeed, 1900 is the plaintext of 2342, since one can easily verify that

$$1900^{17} \bmod 2773 = 2342.$$

So, in this example we did not use any of the trap-door information $(d, p, q, \phi(N))$ to break C.

Remark 6.8.1. The above example actually shows an interesting fact that

$$\begin{aligned} M &\equiv 2342^{157} \\ &\equiv 2342^{8115282064127262599192246347548867233436266l535406513} \\ &\equiv 1900 \pmod{2773} \end{aligned}$$

Remark 6.8.2. Suppose e has order r in the multiplicative group modulo $\lambda(N)$. Then $e^r \equiv 1 \pmod{\lambda(N)}$, so $M^{e^r} \equiv M \pmod{N}$. This is just the rth iterate of the encryption of M. Thus, we must ensure that r is large.

6.9 Chapter Notes and Further Readings

This chapter discussed some cryptanalytic attacks on RSA based on some elementary number-theoretic properties and/or the improper uses of the RSA. Clearly, if the message space $\mathcal{M}$ is small or predictable, the attacker can decrypt the ciphertext C by computing all possible $C' \equiv M'^2 \pmod{N}$ to see if a $C' = C$ can be found. If so, then the M' is the required M. Hence decrypt the ciphertext. This is called the forward search attack. However, this attack will not work if the original plaintext M is appended by random bit string prior to encryption; the random bit string should be independently generated for each encryption, even for the same message M. This process of appending a random string to the original message is known as the message salting. Most of the attacks discussed in this chapter will not be valid if the RSA encryption is used and implemented properly. Thus, these attacks essentially just illustrate the pitfalls to be avoided when implementing and using RSA. The following references provide discussions and/or overviews of the known attacks on RSA from 1978 to present time: Rivest [263], Boneh [37], Coron [83] Countinho [88], Kaliski and Robshaw [162], Koblitz and Menezes [176], and Mollin [217].

The computational equivalence of computing d and factoring N, and computing $\phi(N)$ and factoring N are well known and discussed in many references, say, e.g., Rivest, Shamir and Adleman [262] and Rivest and Kaliski [264]. The common modulus attacks are due to DeLaurentis [96] and Simons [301]. The fixed point attack was first studied by Simons and Morris [300]. Gysin and Seberry in [133] generalized the fixed point attack to the cases without the knowledge of e and C, and hence these attacks can be used as factoring algorithms; They also translated their attacks to elliptic curves. A generalized birthday attack on RSA was also studied by Girault et al proposed in [120].

It is also interesting to note a recent paper by Coron and May [85] that computing d and factoring N are deterministic polynomial-time equivalent. That is.

$$\{d\} \overset{\mathcal{P}}{\Longleftrightarrow} \mathrm{IFP}(N).$$

Previously it is only known [262] that

$$\mathrm{IFP}(N) \overset{\mathcal{P}}{\Longrightarrow} \{d\}, \quad \{d\} \overset{\mathcal{RP}}{\Longrightarrow} \mathrm{IFP}(N).$$

However, this does not imply that factoring N is computationally equivalent to breaking RSA, since breaking RSA may be easier than factoring N, although no proof has been given.

7. Public Exponent Attacks

How can you use one thing's impossibility to make another thing possible that was otherwise impossible too? If we pause to think about it, we would expect nothing of any value to come of negative in algorithmics, except in helping prevent people from wasting time trying to do things that can't be done. Nevertheless, problems for which we have no good solutions are crucial here; in fact, if they turn out to have good solutions we are in big trouble!

DAVID HAREL
Professor of Computer Science, Weizmann Institute of Science

7.1 Introduction

In order to improve the efficiency of the RSA encryption:

$$C \equiv M^e \pmod{N}$$

it is desirable to use a small public (encryption) exponent e such as $e = 3 = (11)_2$ or $e = 17 = (10001)_2$:

$$C \equiv M^3 \pmod{N} \quad \text{or} \quad C \equiv M^7 \pmod{N}.$$

By using the repeated squaring method (Algorithm 1.3.5, starting from the most significant bit), it only requires one modular squaring and one modular multiplication for $e = 3$, and four modular squaring and one modular multiplication for $e = 17$ as follows:

Bits	Operations	Results	Bits	Operations	Results
1	M	M	1	M	M
1	$M^2 \cdot M$	M^3	0	M^2	M^2
			0	$(M^2)^2$	M^4
			0	$((M^2)^2)^2$	M^8
			1	$(((M^2)^2)^2)^2 \cdot M$	M^{17}

However, as the public exponent e and the private exponent d are a pair of invertible elements modulo $\phi(N)$. That is,

$$ed \equiv 1 \pmod{\phi(N)}.$$

Thus, in theory, once the public-key (e, N) is given, d is completely determined:

$$d \equiv 1/e \pmod{\phi(N)}.$$

Of course, the existence of d in the above formula does not mean d can be found efficiently, just the same as that the existence and uniqueness of prime factorization of N does not imply the prime factorization of N can be found efficiently. There are apparently at least two ways to find d:

(1) If $\phi(N)$ is given, or alternatively, if the prime factorization of $N = pq$ is known, then d can be computed by using extended Euclid's algorithm to the linear Diophantine equation $ed - k\phi(N) = 1$.

(2) Even if $\phi(N)$ or the prime factorization of N is unknown, it may still be possible to find d from the given public-key (e, N) in polynomial-time, provided that an improper use of RSA can be discovered, for example, if e is chosen to be small in some circumstances. That is,

$$\{e, N\} \xrightarrow[\text{maybe easy when } e \text{ small}]{\text{find}} \{d\}.$$

In this chapter, we shall study some of these cryptanalytic attacks of the second category.

7.2 A Theorem of Coppersmith

The Coppersmith theorem of 1997 (see [76], [77], [78], [79], [80] and [146]) is one of the most important results in cryptanalytic attacks on RSA, and has a great impact on the attacks on short RSA public exponent e. In fact, the most powerful attacks on short RSA public exponent e are based on the theorem.

Let N be a large odd composite. Let $p(x)$ be a univariate, monic and reducible modular polynomial:

$$p(x) = x^\delta + a_{\delta-1}x^{\delta-1} + \cdots + a_1 x + a_0 \in \mathbb{Z}[x],$$

where $\delta \in \mathbb{Z}^+$ is the degree of the polynomial. For a suitable bound B, we wish to find all small integer roots x_0 such that $|x_0| \leq B$ and $p(x_0) \equiv 0 \pmod{N}$. Such integers of x_0 are called *small roots* or *small solutions* of the univariate modular polynomial equation $p(x)$.

Theorem 7.2.1 (Coppersmith). Let N be a large odd composite. Let $p(x)$ be a univariate, monic and reducible modular polynomial:

$$p(x) = x^{\delta} + a_{\delta-1}x^{\delta-1} + \cdots + a_1 x + a_0 \in \mathbb{Z}[x], \tag{7.1}$$

where $\delta \in \mathbb{Z}^+$ is the degree of the polynomial. Let x_0 be solution to the modular polynomial equation, defined in (7.1):

$$p(x) \equiv 0 \pmod{N}. \tag{7.2}$$

Then if $|x_0| \leq N^{1/\delta}$, x_0 can be computed in time polynomial in δ and $\ln(N)$.

The main technique in the proof of the theorem is the non-trivial use of the lattice basis reduction algorithm, called LLL and invented by Lenstra, Lenstra and Lovász in 1982 [188]; in its simplest non-technical language, the modular polynomial equation of degree δ

$$p(x) \equiv 0 \pmod{N}$$

was transformed to an exact, non-congruence, polynomial equation

$$r(x) = 0$$

by using a lattice of dimension $\delta + 1$, then just solve $r(x)$ for solutions. Solving a congruence polynomial equation requires to factor N, which is hard, whereas solving an exact non-congruence polynomial equation can be done by numerical procedures such as the Newton's method, which is not so hard. Normally, when applying LLL, the value for δ must be small enough so that LLL can be run in reasonable amount of time. But the significance of Coppersmith theorem is that it uses higher dimension lattices to look for all small integer solutions/roots (if they exist) x_0 of the polynomial equation $p(x) \equiv 0 \pmod{N}$, and if $|x_0| \leq N^{1/2}$, then x_0 can be found in polynomial-time. Note that soon after Coppersmith proposed his method in 1996, Howgrave-Graham in 1997 [146] proposed a a new method for finding all small integer roots, $|x_0| \leq N^{1/\delta}$, of Equation (7.1), hence, a new proof of the Coppersmith theorem 7.2.1. Although by the general result of dual lattices, the two methods are equivalent, Howgrave-Graham's is computationally more efficient than Coppersmith's. Howgrave-Graham has specifically presented a nice example of computing a small root of the modular polynomial equation

$$p(x) \equiv x^2 + 14x + 19 \equiv 0 \pmod{35}$$

as follows: First form the 6×6 matrix (in Coppersmith's method, it would need to form a 10×10 matrix) M:

$$M = \begin{pmatrix} 1225 & & & & & \\ 0 & 1225 \cdot 2 & & & 0 & \\ 665 & 490 \cdot 2 & 35 \cdot 2^2 & & & \\ 0 & 665 \cdot 2 & 490 \cdot 2^2 & 35 \cdot 2^3 & & \\ 361 & 532 \cdot 2 & 234 \cdot 2^2 & 28 \cdot 2^3 & 2^4 & \\ 0 & 361 \cdot 2 & 532 \cdot 2^2 & 234 \cdot 2^3 & 28 \cdot 2^4 & 2^5 \end{pmatrix}.$$

Using LLL, M is reduced to B:

$$B = \begin{pmatrix} 3 & 8\cdot 2 & -24\cdot 2 & -8\cdot 2^3 & -1\cdot 2^4 & 2\cdot 2^5 \\ 49 & 50\cdot 2 & 0 & 20\cdot 2^3 & 0 & 2\cdot 2^5 \\ 115 & -83\cdot 2 & 4\cdot 2^2 & 13\cdot 2^3 & 6\cdot 2^4 & 2\cdot 2^5 \\ 61 & 16\cdot 2 & 37\cdot 2^2 & -16\cdot 2^3 & 3\cdot 2^4 & 4\cdot 2^5 \\ 21 & -37\cdot 2 & -14\cdot 2^2 & 2\cdot 2^3 & 14\cdot 2^4 & -4\cdot 2^5 \\ -201 & 4\cdot 2 & 33\cdot 2^2 & -4\cdot 2^3 & -3\cdot 2^4 & 1\cdot 2^5 \end{pmatrix}$$

where $B = HM$ with

$$H = \begin{pmatrix} 70 & 46 & -98 & 32 & -57 & 2 \\ 73 & 48 & -104 & 32 & -56 & 2 \\ 55 & 36 & -74 & 27 & -50 & 2 \\ 125 & 82 & -171 & 60 & -109 & 4 \\ -175 & -115 & 254 & -74 & 126 & -4 \\ 41 & 27 & -59 & 18 & -31 & 1 \end{pmatrix}.$$

Dividing the entries of b_1 by $2^0, 2^1, 2^2, 2^3, 2^4, 2^5$ gives the required polynomial

$$r(x) = 2x^5 - x^4 - 8x^3 - 24x^2 + 8x + 3.$$

Solving this polynomial over $\mathbb{Z}$ gives the solution $x_0 = 3$. Interested readers are suggested to consult Howgrave-Graham's original paper in [146].

Coppersmith's theorem has many novel applications, mostly in cryptology, for example, one of them is in the proof of the result in Theorem 6.5.2:

$$\text{IFP}(N) \overset{\mathcal{P}}{\Longleftrightarrow} \text{RSA}(d).$$

Since we are concerned with cryptanalytic attacks on RSA the most interesting aspect is that the finding of small solutions to Equation (7.1) can lead to various attacks on short public exponent e (some of them will be discussed in this and the next sections). Possibly because of this reason, Lenstra and Verheul suggested on page 265 in [189] that

> *Based on recent results in this area the public exponent for RSA must be sufficiently large. Values such as* 3 *and* 17 *can no longer be recommended, but commonly used values such as* $2^{16} + 1 = 65537$ *still seem to be fine. If one prefers to stay on the safe side one may select an odd 32-bit or 64-bit public exponent at random.*

7.3 Short e Attacks for Same Messages

This section discusses an efficient attack for small e. Let $e = 3$, the smallest possible encryption exponent in the RSA encryption. Suppose that the same massage M was sent to three different recipients using $\{(e, N_1), (e, N_2), (e, N_3)\}$, where $\gcd(N_i, N_j) = 1$ for $i, j = 1, 2, 3$ with $i \neq j$, and $M < N_i$. Then, M can be recovered very efficiently. That is,

$$\{e, C_1, C_2, C_3, N_1, N_2, N_3\} \xrightarrow[\text{easy}]{\text{find}} \{M\}. \tag{7.3}$$

Theorem 7.3.1. Let $e_1 = e_2 = e_3 = 3$ and

$$\begin{cases} C_1 \equiv M^3 \pmod{N_1} \\ C_2 \equiv M^3 \pmod{N_2} \\ C_3 \equiv M^3 \pmod{N_3} \end{cases}$$

Applying the Chinese Remainder Theorem to the system of congruences

$$\begin{cases} M^3 \equiv C_1 \pmod{N_1} \\ M^3 \equiv C_2 \pmod{N_2} \\ M^3 \equiv C_3 \pmod{N_3} \end{cases}$$

we get

$$\begin{aligned} M^3 \quad \equiv \quad & C_1 \cdot (N_2 \cdot N_3) \cdot ((N_2 N_3)^{-1} \bmod N_1) + \\ & C_2 \cdot (N_1 \cdot N_3) \cdot ((N_1 N_3)^{-1} \bmod N_2) + \\ & C_3 \cdot (N_1 \cdot N_2) \cdot ((N_1 N_2)^{-1} \bmod N_3) \pmod{N_1 \cdot N_2 \cdot N_3}. \end{aligned}$$

Thus

$$M \equiv M^{1/3}.$$

Proof. Result follows immediately from the Chinese Remainder Theorem. □

Remark 7.3.1. It is easy to compute the cubic root of x, $x^{1/3}$, although the modular cubic root, $x^{1/3} \bmod N$ with N a large composite, is hard to compute, just the same as the modular square root problem, it is easy to compute $x^{1/2}$ but it is hard to compute $x^{1/2} \bmod N$ with N a large composite.

Example 7.3.1. Let

$$\begin{array}{rcl}
e & = & 3, \\
N_1 & = & 16074226778088739019349815403643227770685367817263534\text{_} \\
 & & 431669799546987155543930704061491272423912871491087467\text{_} \\
 & & 2334376248916284 1829, \\
N_2 & = & 114381625757888867669235779976146612010218296721242362\text{_} \\
 & & 562561842935706935245733897830597123563958705058989075\text{_} \\
 & & 147599290026879543541, \\
N_3 & = & 180708208868740480595165616440590556627810251676940134\text{_} \\
 & & 917012702145005666254024404838734112759081230337178188\text{_} \\
 & & 796656318201321488055 7, \\
M & = & 200805001301070903002315180419000118050019172105011309\text{_} \\
 & & 190800151919090618010705, \\
C_1 & = & M^3 \bmod N_1 \\
 & = & 729191335863603524094702563737402819685723566000524152\text{_} \\
 & & 246303124155589561950490681566138962049778679086940652\text{_} \\
 & & 7191986148822210607 \\
C_2 & = & M^3 \bmod N_2 \\
 & = & 148003742145067300325437785581190327071346176609583779\text{_} \\
 & & 427938716412300234648347175606574700715211712522069717\text{_} \\
 & & 4231808422558905310 7 \\
C_3 & = & M^3 \bmod N_3 \\
 & = & 166741721328924021197326007964121772980620672588921670\text{_} \\
 & & 786680897106877194697211658188467455388280375327727290\text{_} \\
 & & 0087640189457978330457
\end{array}$$

Now suppose Eve knows $(e, C_1, C_2, C_3, N_1, N_2, N_3)$ and wants to find M. She solves the following system of congruences by using the Chinese Remainder Theorem:

$$\begin{cases} M^3 \equiv C_1 \pmod{N_1} \\ M^3 \equiv C_2 \pmod{N_2} \\ M^3 \equiv C_3 \pmod{N_3} \end{cases}$$

to obtain

$$
\begin{aligned}
M^3 = {} & 80969894940479982517531882959928080070140488956432988880473\_ \\
& 59014720253220847686394738491562155963332021601075233618284 26\_ \\
& 25926133135704100964773441035273944310174245220397589398715 10\_ \\
& 2375210975722643202287023918357565845861111152625
\end{aligned}
$$

Now the cryptanalyst, Eve, can easily perform the cubic root of M and gets:

$$
\begin{aligned}
M \quad = \quad & 200805001301070903002315180419000118050019172105011309190 8\_ \\
& 0015191909061801070 5.
\end{aligned}
$$

As can be seen, Eve can break the RSA encryption without factoring N and without using any knowledge of the trap-door information $\{d, p, q, \phi(N)\}$.

Remark 7.3.2. The potential insecure problem of using low public-key exponent e, particularly for $e = 3$, as discussed in this section, does not mean RSA M^3 encryption is not secure. Knuth proposed a challenge RSA M^3 encryption problem in his famous and widely distributed book [169] (page 417); he rated the difficulty of this problem as M50, the hardest class of mathematics problems in his book. Let $C = (M_1^3 \bmod N, M_2^3 \bmod N)$ be as follows:
(687502836437089289878995350604407990716898140258583443035535588237479 27
108009029304963056665126811233405627433261214282318720373118151963944261
656899892436827122751237714587973722992041257530236659548756413 82171,
713013988616927464542046650358646224728216664013755778567223219797011593
220849557864249703775331317377532696534879739201868887567829519032681632
688812750060251822388446286615758360493162805668669968333451929 4663)
and N is a 211-digit composite with two prime factors:
779030228851015954236247565470557836248576762097398394108440222213572872
511709998585048387648131944340510932265136815168574119934775586854274094
225644500087912723258574933706185395834027843405820888108548507873 7.
Find the plaintext M_1 and M_2. This challenge has been opened since 1997.

To defend the short public exponent e attack, it should never send the same message several times even if different moduli are used for different encryption processes. However, if we increase $e = 3$ to $e = 2^{16} + 1 = 65537$ and do the encryption as follows:

$$
\begin{cases}
C_1 \equiv M^{65537} \pmod{N_1} \\
C_2 \equiv M^{65537} \pmod{N_2} \\
C_3 \equiv M^{65537} \pmod{N_3}
\end{cases}
$$

then, M will be difficult to recover from C_1, C_2, C_3. To recover M, the attacker needs to receive 65537 ciphertexts, which is almost impossible, since essentially no-one will send a same message M to 65537 different recipients. On the other hand, it is still very efficient as $e = 3$, since $e = 2^{16} + 1 =$

$65537 = (10000000000000001)_2$, the encryption requires only 17 modular squarings and two modular multiplications using Algorithm 1.3.5.

Now we consider a general case (linear related plaintexts):

$$\begin{cases} C_1 \equiv (a_1 M + b_1)^e \pmod{N_1} \\ C_2 \equiv (a_2 M + b_2)^e \pmod{N_2} \\ \quad \vdots \\ C_k \equiv (a_k M + b_k)^e \pmod{N_3} \end{cases} \tag{7.4}$$

where $k \geq e$ and $\gcd(N_i, N_j) = 1$ for $i, j = 1, 2, \cdots, k$ with $i \neq j$ and $M < \min(N_i)$, and we ask *is it possible to recover M from $C_1, C_2, \cdots, C_k$ in polynomial-time*? The answer is *yes* if the similar (not exactly the same) messages is at least seven [136]. We can consider an even more general case:

$$\begin{cases} p_1(x) \equiv 0 \pmod{N_1} \\ p_2(x) \equiv 0 \pmod{N_2} \\ \quad \vdots \\ p_k(x) \equiv 0 \pmod{N_k} \end{cases} \tag{7.5}$$

where each $p_i(x)$ for $i = 1, 2, \cdots, k$ is a polynomials of degree δ, and N_i are distinct pairwise relatively prime numbers with $x < \min(N_i)$, we ask *is it possible to find all solutions x to (7.5) in polynomial-time*? The answer to this question is again *yes* if $k > \delta(\delta + 1)/2$ provided $\min(N_i) > 2^{\delta^2}$ [136]. Moreover, Hastäd showed that

[1] Sending linearly related messages using RSA with short public exponent e is not secure. If more than $e(e+1)/2$ such messages are sent it is possible for a cryptanalyst to recover the messages in polynomial-time provided that the moduli $N_i > 2^{(e+2)(e+1)/4}(e+1)^{e+1}$.

[2] Sending linearly related messages using Rabin cryptosystem which uses the encryption function $x \mapsto x^2 \pmod{N}$ is not secure. If three such messages are sent it is possible for a cryptanalyst to recover the message in polynomial-time.

Again, similar to Coppersmith and Howgrave-Graham, Hastäd uses LLL to obtain the above results. As a consequence, using small e, e.g., $e = 3, 17$, as public exponent is not a good choice for application of RSA in a large network system. However, this attack may be thwarted by using some salting and padding processes such as OAEP [114] or the one discussed in [87], prior to encryption.

The above discussed attack on short e for same (or similar) messages is known as Hastäd's *broadcasting attack*. A stronger version of Hastäd's attack was proved by Boneh [37]:

Theorem 7.3.2. Let $N_1, N_2, \cdots, N_k$ be pairwise relatively prime with $N_1 \leq N_i$ for all $i = 1, 2, \cdots, k$. Let $p_i(x) \in \mathbb{Z}_{N_i}[x]$ for $i = 1, 2, \cdots, k$ be the k

polynomials with maximum degree δ. Suppose there exists a unique $M < N_1$ such that $p_i \equiv 0 \pmod{N_i}$ for all $i = 1, 2, \cdots, k$ and $k \leq \delta$, then M can be computed in polynomial-time.

Example 7.3.2 (Mollin [217]). Suppose that Bob is communicating with recipient $R_1, R_2, \cdots, R_k$, and the public exponent $e < k$. Suppose now Bob wants to send the same message M to all the k users. Bob performs:

[1] He first pads the message M for each recipient R_i as follows (to increase the security, he shall randomize his padding process each time):

$$p_i(M) = 2^t i + M$$

where t is the bit length of M.

[2] He then sends

$$p_i^e(M) \equiv (2^t i + M)^e \pmod{N_i}$$

for $i = 1, 2, \cdots, k$ to the k recipients.

Then if the cryptanalyst, Eve, intercepts more than e of these $p_i^e(M)$ for $i = 1, 2, \cdots, k$ messages, she can recover M in polynomial-time by Theorem 7.3.2.

7.4 Short e Attacks for Related Messages

In the previous section, we considered attacks on short RSA public exponent e encryption for the same (or the similar) message M. In this section, we consider attacks on short e encryption for the related messages using the same modulus, due to Coppersmith, Franklin and Reiter [81].

Theorem 7.4.1. Let M_1 and M_2 are messages related to the following known relation:

$$M_2 = \alpha M_1 + \beta. \tag{7.6}$$

Suppose the RSA M^3 encryption uses the same modulus N as follows:

$$\left.\begin{array}{l} C_1 \equiv M_1^3 \pmod{N}, \\ C_2 \equiv (\alpha M_1 + \beta)^3 \pmod{N}. \end{array}\right\} \tag{7.7}$$

Then

$$\begin{aligned} M_1 &\equiv \frac{\beta(C_2 + 2\alpha^3 C_1 - \beta^3)}{\alpha(C_2 - \alpha^3 C_1 + 2\beta^3)} \\ &\equiv \frac{3\alpha^3\beta M_1^3 + 3\alpha^2\beta^2 M_1^2 + 3\alpha\beta^3 M_1}{3\alpha^3\beta M_1^2 + 3\alpha^2\beta^2 M_1 + 3\alpha\beta^3} \pmod{N}. \end{aligned} \tag{7.8}$$

Proof. Result follows by the substitution of C_1 and C_2 in (7.7) and (7.8). □

Corollary 7.4.1. If we let $M_1 = M$, $M_2 = M + 1$, and $\alpha = \beta = 1$, then (7.8) reduces to the following special case:

$$\begin{aligned} M &\equiv \frac{(M+1)^3 + 2M^3 - 1}{(M+1)^3 - M3 + 2} \\ &\equiv \frac{3M^3 + 3M^2 + 3M}{3M^2 + 3M + 3} \pmod{N}. \end{aligned} \tag{7.9}$$

Example 7.4.1. Give the relation $M_2 = \alpha M_1 + \beta$ and let

$$\begin{aligned} \alpha &= 3 \\ \beta &= 5 \\ N &= 779030228851015954236247565470557836248576762097398394\_ \\ &\quad 108440222213572872511709998585048387648131944340510932\_ \\ &\quad 265136815168574119934775586854274094225644500087912723\_ \\ &\quad 258574933706185395834027843405820888108548507 8737 \\ C_1 &= 132057584044937409231208389323398996878812486949811558\_ \\ &\quad 724214983072091380989054308161277959733824865068687594\_ \\ &\quad 213139826622055543700074552293693503940351187203266740\_ \\ &\quad 911056806170880679978462212228231292575333924006 \\ C_2 &= 356555476921331004924262651173177291572793714764491209\_ \\ &\quad 099736209686208403812308104374465892532943045181265208\_ \\ &\quad 185871222090592859132787427488883517622574112296645299\_ \\ &\quad 299833541045392916173339389220473000267483895 5287 \end{aligned}$$

Then

$$\begin{aligned} M_1 &\equiv \frac{5(C_2 + 2 \cdot 3^3 \cdot C_1 - 5^3)}{3(C_2 - 3^3 \cdot C_1 + 2 \cdot 5^3)} \\ &\equiv 200805001301070903002315180419000118050019172105011309\_ \\ &\quad 190800151919090618010705 \pmod{N} \\ M_2 &\equiv 3M_1 + 5 \pmod{N} \\ &\equiv 602415003903212709006945541257000354150057516315033927\_ \\ &\quad 572400455757271854032120 \pmod{N} \end{aligned}$$

It is easy to verify that M_1 and M_2 are the correct plaintexts of C_1 and C_2, since

$$
\begin{aligned}
C_1 &\equiv M_1^3 \bmod N \\
&= 13205758404493740923120838932339899687881248694981155\_ \\
&\quad 72421498307209138098905430816127795973382486506868759 4\_ \\
&\quad 21313982662205554370007455229369350394035118720326674 0\_ \\
&\quad 911056806170880679978462212228231292575333924006 \\
C_2 &\equiv M_2^3 \bmod N \\
&= 356555476921331004924262651173177291572793714764491209\_ \\
&\quad 099736209686208403812308104374465892532943045181265208\_ \\
&\quad 185871222090592859132787427488883517622574112296645299\_ \\
&\quad 2998335410453929161733393892204730002674838955287
\end{aligned}
$$

as required. Note that to recover M_1 and M_2 from C_1 and C_2 in the present case, it is no need to factor N, the 211-digit modulus, and also no need to find d.

Remarkably enough, when $e > 3$, the above attack on RSA will still work.

Theorem 7.4.2. Let $e = 5$, and set

$$
\left.
\begin{aligned}
C_1 &\equiv M^5 \pmod{N}, \\
C_2 &\equiv (M+1)^5 \pmod{N}.
\end{aligned}
\right\} \tag{7.10}
$$

Then

$$
\left.
\begin{aligned}
G &\equiv C_2^3 - 3C_1C_2^2 + 3C_1^2C_2 - C_1^3 + 37C_2^2 + 176C_1C_2 + 37C_1^2 \\
&\quad +73C_2 - 73C_1 + 14 \pmod{N} \\
F &\equiv M \cdot G \equiv 2C_2^3 - C_1C_2^2 - -4C_1^2C_2 + 3C_1^3 + 14C_2^2 - 88C_1C_2 \\
&\quad -51C_1^2 - 9C_2 + 64C_1 - 7 \pmod{N}.
\end{aligned}
\right\} \tag{7.11}
$$

Hence

$$
M \equiv F/G \pmod{N}. \tag{7.12}
$$

Proof. Result follows by substituting C_1 and C_2 in (7.10) to G and F in (7.11), respectively. □

Example 7.4.2. Let again

$$
N = p \cdot q = 179424797 \cdot 197747359 = 35480779745861123.
$$

Suppose also we have

$$\begin{aligned}
M &= 19443855347586362 \\
C_1 &\equiv M_1^5 \\
&\equiv 19443855347586362^5 \\
&\equiv 18796237015415790 \pmod{35480779745861123} \\
C_2 &\equiv (M+1)^5 \\
&\equiv (19443855347586362+1)^5 \\
&\equiv 7290180156009373 \pmod{35480779745861123}.
\end{aligned}$$

Now suppose we are given

$$\begin{aligned}
N &= 35480779745861123 \\
C_1 &= 18796237015415790 \\
C_2 &= 7290180156009373
\end{aligned}$$

then by applying the attack, we can recover M and $(M+1)^5$ from (N, C_1, C_2) as follows:

$$\begin{aligned}
G &= C_2^3 - 3C_1C_2^2 + 3C_1^2C_2 - C_1^3 + 37C_2^2 + 176C_1C_2 + 37C_1^2 \\
&\quad +73C_2 - 73C_1 + 14 \\
&= -1523279324842988466954772791427559863314856282747
\end{aligned}$$

$$\begin{aligned}
F &= M \cdot G \\
&= 2C_2^3 - C_1C_2^2 - 4C_1^2C_2 + 3C_1^3 + 14C_2^2 - 88C_1C_2 - 51C_1^2 \\
&\quad -9C_2 + 64C_1 - 7 \\
&= 939554883348046995835655578034119536835154607906
\end{aligned}$$

$$\begin{aligned}
M &\equiv F/G \\
&\equiv \frac{939554883348046995835655578034119536835154607906 6}{-1523279324842988466954772791427559863314856282747} \\
&\equiv 19443855347586362 \pmod{35480779745861123}.
\end{aligned}$$

So, we have done, and everything is correct, as expected.

Clearly, when e becomes large, the algebraic formulas for computing the two polynomials G and F in C_1 and C_2 will be complicated and require $\mathcal{O}(e^2)$ coefficients. Fortunately, there is a simple and direct method for arbitrary e.

Proposition 7.4.1. Let $x = M$ denote the unknown message m, then x satisfies:

$$\left.\begin{aligned}
x^e - C_1 \equiv 0 \pmod{N} \\
(x+1)^e - C_2 \equiv 0 \pmod{N}
\end{aligned}\right\} \tag{7.13}$$

Then

$$\gcd(x^e - C_1, (x+1)^5 - C_2) \in \mathbb{Z}_N[x], \tag{7.14}$$

which will almost produce the linear polynomial $x - M$.

The attack applies to any values of e. By using some sophisticated technique in the theory of continued fractions [313], the computation of gcd of two polynomials of degree e over ring $\mathbb{Z}_N[x]$ can be performed in time $\mathcal{O}((e \log e)^2)$. Thus, the attack may be practical for all e with length up to 32 bits. Hence, the attack will be very efficient for $e = 2^{16} + 1 = 65537$.

Example 7.4.3. Let

$$\begin{aligned}
M_2 &= 2M_1 + 3 \pmod N \\
e &= 5 \\
N &= 35480779745861123 = 179424797 \cdot 197747359 \\
C_1 &= E(M_1) = 18796237015415790 \\
C_2 &= E(M_2) = 7290180156009373
\end{aligned}$$

Suppose the authorized people know M_1, M_2, C_1, C_2 as follows

$$\begin{aligned}
M_1 &= 19443855347586362 \\
M_2 &= 2M_1 + 3 \\
&\equiv 2 \cdot 19443855347586362 + 3 \\
&\equiv 3406930949311604 \pmod{35480779745861123} \\
C_1 &\equiv M_1^5 \\
&\equiv 19443855347586362^5 \\
&\equiv 18796237015415790 \pmod{35480779745861123} \\
C_2 &\equiv (2M+3)^5 \\
&\equiv (2 \cdot 19443855347586362 + 3)^5 \\
&\equiv 30744837049745686 \pmod{35480779745861123}
\end{aligned}$$

The attacker, however, only knows (N, e, C_1, C_2) but he wants to know M_1 and M_2. By Proposition (7.4.1), he can recover the two related unknown plain-texts (M_1, M_2) from the known ciphertexts (C_1, C_2) by computing the greatest common divisor of two polynomials over the integer ring $\mathbb{Z}_N[x]$, where $N = pq$ with p, q primes:

$$\gcd(z^5 - 18796237015415790, (2z+3)^5 - 30744837049745686) \pmod N.$$

Of course, he cannot directly find the gcd, but there is a very efficient indirect method to compute the gcd, as follows:

[1] Compute the gcd over $\mathbb{Z}_p$ and $\mathbb{Z}_q$:

$$\begin{aligned}
&\gcd(x^5 - 18796237015415790, (2x+3)^5 - 30744837049745686)\\
&\quad \equiv 97067290 + z \pmod{179424797}\\
&\gcd(x^5 - 18796237015415790, (2x+3)^5 - 30744837049745686)\\
&\quad \equiv 179461606 + z \pmod{197747359}
\end{aligned}$$

[2] Use the Chinese Remainder Theorem to solve the following system of congruence equations:

$$\begin{cases} y \equiv 97067290 + x \pmod{179424797} \\ y \equiv 179461606 + x \pmod{197747359} \end{cases}$$

and get

$$\begin{aligned}
&\gcd(x^5 - 18796237015415790, (2x+3)^5 - 30744837049745686)\\
&\quad \equiv 16036924398274761 + x\\
&\quad \equiv x - 19443855347586362 \pmod{35480779745861123}.
\end{aligned}$$

So, he concludes that

$$\begin{aligned}
M_1 &= 19443855347586362\\
M_2 &= 2M_1 + 3\\
&= 3406930949311604.
\end{aligned}$$

Thus, he successfully and efficiently recovers the original plaintexts (M_1, M_2).

In [81], Coppersmith, et al even considered the case when the k messages $M_1, M_2, \cdots, M_k$ are related by a polynomial $p(M_1, M_2, \cdots, M_k)$, and we know the ciphertexts $C_1, C_2, \cdots, C_k$ and the coefficients of the polynomial p.

Proposition 7.4.2. Suppose there are k messages $m_1, m_2, \cdots, m_k$, related by a polynomial $p(m_1, m_2, \cdots, m_k)$ with degree δ. Suppose also that we know the ciphertexts $C_i \equiv M_i^e \pmod N$ and the coefficients of the polynomial p. By substituting the variables Z_i for the unknown message M_i, the following $k+1$ polynomials can be obtained:

$$\begin{aligned}
p_0(x_1, x_2, \cdots, x_k) &\equiv p(x_1, x_2, \cdots, x_k) &&\equiv 0 \pmod N\\
p_1(x_1) &\equiv x_1^e - C_1 &&\equiv 0 \pmod N\\
p_2(x_2) &\equiv x_2^e - C_2 &&\equiv 0 \pmod N\\
&\vdots\\
p_k(x_k) &\equiv x_k^e - C_k &&\equiv 0 \pmod N
\end{aligned}$$

Then, by computing the Gröbner base [334]

$$\text{Grobner}([P_0, P_1, P_2, \cdots, P_k])$$

the final result will thus generally be generated:

$$[x_1 - M_1, x_2 - M_2, \cdots, x_k - M_k].$$

Interested readers are suggested to consult the original paper [81] for more information and particularly for the implications of these attacks to the security of some cryptographic protocols.

The short e attacks on related messages discussed above might seem to be rather artificial, since in the real world situation, two plaintexts may not be related at all. A strengthened version of the attack on random padded and related messages, known as *short pad attacks*, due to Coppersmith [78], may be described as follows. Suppose Bob want to send a message M to Alice. He randomly pads M and transmits the resulting ciphertext. Eve, the cryptanalyst, now has two ciphertexts of the same message using two different random pads. Coppersmith shows in the following theorem that although Eve does not know the pads Bob used, she can recover the plaintext in polynomial-time.

Theorem 7.4.3 (Coppersmith). Let N be a β bit RSA modulus, and e the public exponent. Set $m = \lfloor N/e^2 \rfloor$. Suppose that $M \in \mathbb{Z}_N^*$ is the plaintext message at most $\beta - m$ bits. Define

$$\begin{aligned} M_1 &\equiv 2^m M + r_1 \\ M_2 &\equiv 2^m M + r_2 \end{aligned}$$

where r_1, r_2 are two distinct integers with $r_1 \neq r_2$ and $0 \leq r_1, r_2 < 2^m$. Then given (N, e) and the ciphertexts C_1, C_2 of M_1, M_2, but it is not given r_1 or r_2, M can be efficiently recovered.

Corollary 7.4.2. Let $e = 3$. Then the short pad attack can be mounted as soon as the pad length is less than 1/9 of the message length.

This is true because if the message length is γ, then

$$\{r_1, r_2\} < 2^{\gamma/9} < 2^{\beta/9} = 2^m.$$

Note that the short pad attack cannot be mounted against the standard moduli sizes if $e = 65537$.

7.5 Lattice Attack for Stereotyped Messages

When you open a new account in a bank, your bank will normally send you a pin number usually in the following form:

Your pin number is ****

or when you open a new computer account in the computing services of your university, you will normally be informed by

Your login is **** and your passwd is ****

Such messages (consisting of a fixed known string and an unknown number/string) are called *stereotyped messages* on page 243 of [78]. Suppose the messages are encrypted by RSA. We will show in this section that if the unknown number **** is chosen to be small, then it can be easily recovered. For the simplicity of the discussion, we consider the first case which is of the form

$$\begin{aligned} M &= B + x, \\ C &\equiv M^3 \pmod{N} \end{aligned}$$

Thus, by solving the modular cubic polynomial equation

$$(B + x)^3 - C \equiv 0 \pmod{N} \tag{7.15}$$

one can recover x and then find the original plaintext message M. The idea is to use Coppersmith's LLL technique (i.e., Theorem 7.2.1) to find small solutions to the modular polynomial equation (7.15).

Theorem 7.5.1. Let the modular cubic polynomial equation be as follows:

$$(B + x)^3 - C \equiv 0 \pmod{N} \tag{7.16}$$

where

$$\begin{aligned} M &= B + x, \\ C &\equiv M^3 \pmod{N} \end{aligned}$$

with $N = pq$ and $e = 3$ the RSA public-key, B a fix known string, x an unknown variable, M the plainttext, and C the cyphertext, respectively. Then, all solutions x_0 (if they exist) to (7.16) can be found in polynomial-time provided that $|x_0| < N^{1/3}$.

The above theorem follows immediately from Theorem 7.2.1.

Remark 7.5.1. As noted in [78], the bound on x_0 depends on N. If x_0 has e.g., 250 bits and N has 512 bits, and $e = 3$, then the present attack will not work, since $x_0 > N^{1/3}$. However, if we increase N to 1024 bits, but still keep x_0 at 250 bits, then these x_0 are vulnerable/recoverable to the attack since $x_0 < N^{1/3}$. The present attack works for when x_0 lies in the least significant bits of the message M. However, it will also work for x_0 lies in the most significant bits of the message M; just multiplying x by a known constant 2^k.

Remark 7.5.2. As mentioned at the beginning of the section, there are types of the stereotyped messages such as

Your pin number is ****

Your login is **** and your passwd is ****

It is interesting to note that the present attack to recover all small solutions x_0 to the univariate modular polynomial:

$$p(x) \equiv (B + x^3) \equiv 0 \pmod{N}$$

where $M = B + x$ and $C \equiv M^3 \pmod{N}$, can be extended to recover small solutions x_0, y_0 to the bivariate modular polynomial:

$$p(x, y) \equiv C - (B + 2^k x + y)^3 \equiv 0 \pmod{N},$$

where $M = B = 2^k x + y$ and $C \equiv M^3 \pmod{N}$, provided that we know or can correctly guess the length of x and y. For more information on this extended attack, see [78].

In what follows, we present a small example to show how to use Coppersmith's LLL reduction technique to find the solution x of the modular polynomial equation.

Example 7.5.1. Let

$$\begin{aligned} N &= pq \\ &= 9545724696579737 \cdot 5757285757575823 \\ &= 54957464841358314276864542898551. \end{aligned}$$

Suppose a Bank wishes to send the following message to Bob:

Your pin no is ****

where $* * **$ denotes a four-digits number. The message is of the form

$$M = B + x$$

where $B = 25152118001609140014150009190000$ and $0 \leq x < 10000$. To stop anyone (e.g., Eve), except Bob, to read the message and to get the Pin Number, the Bank actually sends the following cipher-text of the original message to Bob

$$\begin{aligned} C &\equiv (B + x)^3 \\ &\equiv 37393323096087665763922106857101 \\ &\quad \pmod{54957464841358314276864542898551} \end{aligned}$$

Eve, the cryptanalyst, wants to know what is x? She can perform the following steps in order to get x:

[1] Form a polynomial $f(x)$ as follows:

$$\begin{aligned} f(x) &\equiv (B+x)^3 - C \\ &\equiv B^3 + 3B^2x + 3Bx^2 + x^3 - C \\ &\equiv x^3 + 3Bx^2 + 3B^2x + (B^3 - C) \\ &\equiv x^3 + a_2x^2 + a_1x + a_0 \pmod{N} \end{aligned}$$

where

$$\begin{aligned} a_2 &\equiv 3B \\ &\equiv 204988891634691057655854846714449 \pmod{N} \\ a_1 &\equiv 3B^2 \\ &\equiv 318450214368646973392092630354 98 \pmod{N} \\ a_0 &\equiv B^3 - C \\ &\equiv 18283973072868139826273442498263 \pmod{N} \end{aligned}$$

[2] Let $y = 10000$ and form a lattice of vectors

$$\begin{aligned} v_1 &= \{N, 0, 0, 0\} \\ v_2 &= \{0, yN, 0, 0\} \\ v_3 &= \{0, 0, y^2N, 0\} \\ v_4 &= \{a_0, ya_1, y^2a_2, y^4\}. \end{aligned}$$

Then use the lattice reduction algorithm LLL to generate a new basis (b_1, b_2, b_3, b_4) as follows:

([-185048941351041096883325629 76,137474691810128112063488610000,
-2412728694725908630150000000 0,2268379855321848000000000000 00],

[-854250031318348859400871 91,8952128526937476800778600000,
-771386873484388142780000000 00,1211988546742050290000000000000],

[-118201022467767601176264834 51,690469193086730083813708 70000,
-118201022467767601176264834 51, 690469193086730083813708 70000],

[539165082686996693140407765499 41, 739792174410258624319670999 0000,
1080331017291638854711200000000, 911950755583134230000000000000])

[3] Eve uses e.g., the basis vector b_1 to form a new polynomial $r(x)$:

$$r(x) = e_3x^3 + e_2x^2 + e_1x + e_0$$

where

$$b_1 = \begin{cases} e_3 = 2268379855321848000000000000 \\ e_2 = -241272869472590863015000000000 \\ e_1 = 137474691810128112063488610000 \\ e_0 = -185048941351041096883325629766 \end{cases}$$

Of course, she can use other vectors b_2, b_3, b_4 in the basis as well, e.g.,

$$b_2 = \begin{cases} e_3 = 1211988546742050290000000000000 \\ e_2 = -771386873484388142780000000000 \\ e_1 = 895212852693747680077860000000 \\ e_0 = -854250031318348859400871911 \end{cases}$$

or

$$b_3 = \begin{cases} e_3 = 1011923264553169260000000000000 \\ e_2 = 1069165561482498492404000000000 \\ e_1 = 690469193086730083813708700000 \\ e_0 = -1182010224677676011762648345 1 \end{cases}$$

or

$$b_4 = \begin{cases} e_3 = 9119507555831342300000000000000 \\ e_2 = 1080331017291638854711200000000 \\ e_1 = 7397921744102586243196709990000 \\ e_0 = 5391650826869966931404077654994 1 \end{cases}$$

[4] Solving the polynomial equation $r(x) = 0$, based on e.g., b_1, Eve gets:

$$x = \begin{cases} 1379 \\ 52492.26074 + 56216.26293i \\ 52492.26074 - 56216.26293i \end{cases}$$

As the only integer solution to the equation $r(x) = 0$ is $x = 1379$, so Eve concludes that Bob's Pin Number is 1379. Eve of course can try other vectors b_2, b_3, b_4 of the basis to find the required x.

7.6 Chapter Notes and Further Reading

To speed-up the computation in the RSA encryption process, it is very necessary for the public exponent e to be small. For example, PEM [19] recommends 3, the RSA Encryption Standard PKCS #1 [271] recommends 3 or 65537, and the ITU-T standard for public key infrastructure X.509 [154] recommends 65537. However, when e is small, M may be computed efficiently from the given information (e, N, C). It is conjectured that breaking RSA is equivalent to factoring the RSA modulus N, but obviously this cannot be true if e is small. Hastäd [136] may be the first to systematically study the small public exponent e attack; he showed, among others, that sending the linearly related messages using RSA with low exponent e, such as $C_i \equiv (a_i M + b_i)^e \pmod{N_i}$ with $i = 1, 2, 3, \cdots, k$, a_i and b_i are known, is insecure; in fact, sending $k \geq e(e+1)/2$ messages enables the eavesdropper to recover M from C_i, provided that the condition $N_i > 2^{(e+1)(e+2)/4}(e+1)^{e+1}$ is satisfied. Remarkably enough, this is also true for the elliptic curve analogues of RSA, say, e.g, the KMOV analogue [180] and the Demytko analogue [98]. For example, Kurosawa et al [182] showed that for $N \geq 2^{1024}$, if $e = 5$ and $k \geq 428$, then M can be computed from $C_1, C_2, \cdots, C_k$ in polynomial-time for both the KMOV analogue and the Demytko analogue for elliptic curves over $\mathbb{Z}_N$ with $N = pq$, where p and q are primes.

The attacks on small public exponent e for polynomially related messages was studied in Coppersmith, et al [81]; it is particularly useful since several cryptographic protocols are based on encryptions of the polynomially related messages. Recent result shows that the public exponent e for RSA must be sufficiently large. Values such as 3 and 17 can no longer be recommended, but commonly used values such as $2^{16} + 1 = 65537$ will still be acceptable. Currently, a random odd 32-bit or 64-bit number is a good choice for e.

Several attacks on RSA encryption with small public exponent e for both same messages and related (linear and polynomial related) messages have been studied in this section. Readers who wish to get more information about the attacks on RSA low public exponent e are suggested to consult the following references for more information: Boneh [37], Brier et al [56], Coppersmith [78], Coppersmith and Franklin et al [81], Coron et al [84], Hastäd ([139] and [136]), Gueron and Seifert [131], Kurosawa et al [182], Lenstra and Verheul [189], Salah, Darwish and Oqeili [273], Sun and Wu [314], and Jacobi and Jacobi [155]. Fouque et al proposed in [113] a power attack on small RSA public exponent, and Kurosawa et al [182] studied attack for elliptic curve RSA of KMOV [180] and Demytko [98].

Crouch and Davenport [91] considered a hypothetical situation in which low public exponent e is used to encrypt IP packets, TCP segments or TCP segments carried in IP packets. For this scenario, they applied the Coppersmith [78] and Howgrave-Graham [146] lattice methods, in conjunction with the TCP/IP protocols, to decrypt the specific packets when they get retransmitted due to a denial-of-service attack on the receiver's side.

It should be noted that although in RSA cryptosystem, (e, N) is public information, it is still a good idea to keep it as secret as possible, since, as we shall discuss later, an adversary may retrieve secret information from the public information ([56] and [131]); this is why most of the security agencies do not disclose their public-key information to the public even if they use the public-key cryptosystems.

8. Private Exponent Attacks

The protection provided by encryption is based on the fact that most people would rather eat liver than do mathematics.

BILL NEUGENT
Cybersecurity Consultant, Virginia

8.1 Introduction

In the previous chapter, we studied some cryptanalytic attacks applied to the case that the RSA public exponent e is small. That is, when e is small in some circumstances, then given (e, N), M can be recovered from C efficiently:

$$\{e, N\} \xrightarrow[e \text{ is small}]{\mathcal{P}} \{M\} \tag{8.1}$$

Thus, it is a good idea to try to avoid the use of small e such as $e = 3, 17$ whenever possible although short e is a common way to speed up the computation of RSA encryption (and RSA signature verification).

The same situation happens for the RSA private exponent d. To speed up the computation in RSA decryption (and RSA signature generation), it is often to use small private exponent d. Unfortunately, as we shall see in this chapter, the use of small private exponent d is not a good idea and can lead to a total break of the RSA system:

$$\{e, N\} \xrightarrow[d \text{ is small}]{\mathcal{P}} \{d\} \tag{8.2}$$

That is, the use of small d is more serious than the use of small e in RSA system. This is because d is more important than any other piece of trap-door information $(p, q, \phi(N))$. Once d is known, then it is an easy task to break RSA. Thus, whenever possible, a larger d should be used.

8.2 Diophantine Attack

In this section, we shall see that if the RSA private exponent d is chosen too small, e.g., $d < N^{0.25}$, then by Weiner's Diophantine attack [324], d can be efficiently recovered (in polynomial-time) from the public exponent e. That is,

$$\{e, N\} \xrightarrow[d\ <\ N^{0.25}]{\mathcal{P}} \{d\} \tag{8.3}$$

Weiner's idea is based on the properties of continued fractions and the idea of Diophantine approximation.

In as early as 1842 Dirichlet showed that:

Theorem 8.2.1. For any real α and any integer $Q > 1$, there exist integers p and q with $0 < q < Q$ such that

$$|q\alpha - p| \le \frac{1}{Q}. \tag{8.4}$$

This can be generalized to:

Corollary 8.2.1. For any irrational α, there exist infinitely many integers p/q with $q > 0$ such that

$$\left|\alpha - \frac{p}{q}\right| < \frac{1}{q^2}. \tag{8.5}$$

Theorem 8.2.2. For any real α, each convergent p/q satisfies

$$\left|\alpha - \frac{p}{q}\right| < \frac{1}{2q^2}. \tag{8.6}$$

Moreover,

$$\frac{p}{q} = \frac{p_i}{q_i}, \text{ for some } i. \tag{8.7}$$

Theorem 8.2.3. Let $N = pq$ with p, q primes and $q < p < 2q$. Let $1 < e, d < \phi(N)$ with $ed \equiv 1 \pmod{\phi(N)}$. If $d < \frac{1}{3}\sqrt[4]{N}$, then d can be computed in polynomial-time.

First we give a lemma relating to a property of continued fraction expansion of a rational number.

Lemma 8.2.1. Suppose that $\gcd(e, N) = \gcd(k, d) = 1$ and

$$\left|\frac{e}{N} - \frac{k}{d}\right| < \frac{1}{2d^2}.$$

Then k/d is one of the convergents of the continued fraction expansion of e/N.

Proof. Let $N = pq$ with p, q prime and $1 < p < 2q$. Let also the private exponent d is small, say, e.g.,

$$d < \frac{1}{3}\sqrt[4]{N}.$$

Then, as we shall show, d will be the denominator of a convergent to the continued fraction expansion of e/N. □

Theorem 8.2.4 (Weiner). Let $N = pq$ with p and q primes such that

$$\begin{cases} q < p < 2q \\ d < \frac{1}{3}\sqrt[4]{N} \end{cases} \tag{8.8}$$

then given e with $ed \equiv 1 \pmod{\phi(N)}$, d can be efficiently calculated.

Proof. Since $ed \equiv 1 \pmod{\phi(N)}$,

$$ed - k\phi(N) = 1$$

for some $k \in \mathbb{Z}$. Therefore

$$\left| \frac{e}{\phi(N)} - \frac{k}{d} \right| = \frac{1}{d\phi(N)}.$$

Since $N = pq > q^2$, we have $q < \sqrt{N}$. Also since $\phi(N) = N - p - q + 1$,

$$0 < N - \phi(N) = p + q - 1 < 2q + q - 1 < 3q < 3\sqrt{N}.$$

Now,

$$\begin{aligned}
\left| \frac{e}{N} - \frac{k}{d} \right| &= \left| \frac{ed - kN}{dN} \right| \\
&= \left| \frac{ed - kN + k\phi(N) - k\phi(N)}{dN} \right| \\
&= \left| \frac{1 - k(N - \phi(N))}{dN} \right| \\
&< \frac{3k\sqrt{N}}{dN} \\
&= \frac{3k}{d\sqrt{N}} \\
&< \frac{1}{2d^2}.
\end{aligned}$$

Thus, by Lemma 8.2.1, k/d must be one of convergents of the simple continued fraction e/N. Therefore, if $d < \frac{1}{3}\sqrt[4]{N}$, the d can be computed via the elementary task of computing a few convergent of e/N, which can be done in polynomial-time. □

Theorem 8.2.4 tells us that the private-key d should be large enough (nearly as many bits as the modulus N); or otherwise, by the properties of continued fractions, the private-key d can be found in time polynomial in the length of the modulus N, and hence decrypt RSA(M).

Example 8.2.1. Suppose that $N = 160523347$ and $e = 60728973$. Then the continued fraction expansion of e/N is as follows:

$$\begin{aligned}\frac{e}{N} &= 0 + \cfrac{1}{2 + \cfrac{1}{1 + \cfrac{1}{1 + \cfrac{1}{1 + \cfrac{1}{4 + \cfrac{1}{12 + \cfrac{1}{102 + \cfrac{1}{1 + \cfrac{1}{1 + \cfrac{1}{2 + \cfrac{1}{3 + \cfrac{1}{2 + \cfrac{1}{2 + \cfrac{1}{36}}}}}}}}}}}}}} \\ &= [0, 2, 1, 1, 1, 4, 12, 102, 1, 1, 2, 3, 2, 2, 36]\end{aligned}$$

and the convergents of the continued fraction are as follows:

$$[0, \frac{1}{2}, \frac{1}{3}, \frac{2}{5}, \frac{3}{8}, \frac{14}{37}, \frac{171}{452}, \frac{17456}{46141}, \frac{17627}{46593}, \frac{35083}{92734}, \frac{87793}{232061}, \frac{298462}{788917},$$

$$\frac{684717}{1809895}, \frac{1667896}{4408707}, \frac{60728973}{160523347}].$$

If condition (8.8) is satisfied, then the unknown fraction k/d is a close approximation to the known fraction of e/N. Lemma 8.2.1 tells us that k/d must be one of the covergents of the continued fraction expansion of e/N. As can be seen, there are 15 such convergents; we need to find the "right" one. There are two methods to find the right value for d.

(1) The first method is to use the following trial-and-error procedure: if k/d is a convergent of e/N, then we compute $\phi(N) = (ed - 1)/k$ and solve the quadratic equation $p^2 - (N - \phi(N) + 1)p + N = 0$, and check if it leads to a factorization of N. If yes, use the factors p and q just found

to compute $d \equiv 1/e \pmod{\phi(N)}$ and to recover M from C. If no, then pick up the next convergent of e/N and try again. The following are the trying process starting from the second convergents:

Suppose $\frac{k}{d} = \frac{1}{2}$, then

$$\phi(N) = \frac{ed-1}{k} = 60728973 \cdot 2 - 1 = 121457945,$$

$$p^2 - (N - \phi(N) + 1)p + N = 0$$

$$\Longrightarrow p = \frac{39065403}{2} - \frac{\sqrt{1526105069459021}}{2}$$

which is impossible, thus $d \neq 2$.

Suppose $\frac{k}{d} = \frac{1}{3}$, then

$$\phi(N) = \frac{ed-1}{k} = 60728973 \cdot 3 - 1 = 182186918,$$

$$p^2 - (N - \phi(N) + 1)p + N = 0$$

$$\Longrightarrow p = -10831785 - \sqrt{117327405762878}$$

which is impossible, thus $d \neq 3$.

Suppose $\frac{k}{d} = \frac{2}{5}$, then

$$\phi(N) = \frac{ed-1}{k} = \frac{60728973 \cdot 5 - 1}{2} = 151822432,$$

$$p^2 - (N - \phi(N) + 1)p + N = 0$$

$$\Longrightarrow p = 4350458 - 3\sqrt{2102924920713}$$

which is impossible, thus $d \neq 5$.

Suppose $\frac{k}{d} = \frac{3}{8}$, then

$$\phi(N) = \frac{ed-1}{k} = \frac{60728973 \cdot 8 - 1}{3} = \frac{485831783}{3},$$

$$p^2 - (N - \phi(N) + 1)p + N = 0$$

$$\Longrightarrow p = -\frac{4261739}{6} - \frac{18156640463629}{6}$$

which is impossible, thus $d \neq 8$.

Suppose $\frac{k}{d} = \frac{14}{37}$, then

$$\phi(N) = \frac{ed-1}{k} = \frac{60728973 \cdot 37 - 1}{14} = 160498000,$$

$$p^2 - (N - \phi(N) + 1)p + N = 0$$

$$\Longrightarrow p = 12347, \;\; q = N/p = 13001.$$

Sucessful, since $N = pq = 12347 \cdot 13001 = 160523347$.

Given p, q, we can now find $d \equiv 1/e \equiv 37 \pmod{(12347-1)(13001-1)}$ easily, and hence, find the plain-text M. Of course, the above procedure to find d by finding (p, q) after the convergents of e/N is known is not the only possible procedure.

(2) The second possible method is to test whether or not $a^{ed} \equiv a \pmod{N}$ for some randomly chosen a, since d should be one of the denominators

$$\{2, 3, 5, 8, 37, 452, 46141, 46593, 92734, 232061, 788917, \\ 1809895, 4408707, 160523347\}$$

of the e/N convergents, but of course we do not know which one. Thus, we can simply test:

$$2^{60728973 \cdot 2} \bmod 160523347 = 137369160 \neq 2$$
$$2^{60728973 \cdot 3} \bmod 160523347 = 93568289 \neq 2$$
$$2^{60728973 \cdot 5} \bmod 160523347 = 73692312 \neq 2$$
$$2^{60728973 \cdot 8} \bmod 160523347 = 30860603 \neq 2$$
$$2^{60728973 \cdot 37} \bmod 160523347 = 2$$

which gives $d = 37$, as required.

Remark 8.2.1. The above attack in fact gives all the trap-door information $\{d, \phi(N), p, q\}$ as follows:

$$\begin{aligned} \phi(N) &= (ed-1)/p_i \\ &= (60728973 \cdot 37 - 1)/14 \\ &= 160498000, \\ \{p, q\} &\Leftarrow x^2 - (N - \phi(N) + 1) + N \\ &\Rightarrow x^2 - (160523347 - 160498000 + 1)x + 160523347 \\ &\Rightarrow \{p, q\} = 12347 \cdot 13001 \end{aligned}$$

Thus, for $N = 160523347$, Wiener's attack works well for

$$d < \frac{\sqrt[4]{N}}{3} \approx 37.52.$$

Example 8.2.2. Let

$$\begin{aligned} N &= 28562942440499 \\ e &= 7502876735617. \end{aligned}$$

Then the continued fraction expansion of e/N is as follows:

$$\frac{e}{N} = [0, 3, 1, 4, 5, 1, 1, 3, 16, 1, 7, 1, 7, 1, 4, 1, 1, 2, 1, 1, 3, 3, 4, 2, 2, 4, 12, 3, 2, 1, 2]$$

and the convergents of the continued fraction are as follows:

$$[0, \frac{1}{3}, \frac{1}{4}, \frac{5}{19}, \frac{26}{99}, \frac{31}{118}, \frac{57}{217}, \frac{202}{769}, \frac{3289}{12521}, \frac{3491}{13290}, \frac{27726}{105551}, \frac{31217}{118841}, \frac{246245}{937438},$$

$$\frac{277462}{1056279}, \frac{1356093}{5162554}, \frac{1633555}{6218833}, \frac{2989648}{11381387}, \frac{7612851}{28981607}, \frac{10602499}{40362994}, \frac{18215350}{69344601},$$

$$\frac{65248549}{248396797}, \frac{213960997}{814534992}, \frac{921092537}{3506536765}, \frac{2056146071}{7827608522}, \frac{5033384679}{19161753809},$$

$$\frac{22189684787}{84474623758}, \frac{271309602123}{1032857238905}, \frac{836118491156}{3183046340473}, \frac{1943546584435}{7398949919851},$$

$$\frac{2779665075591}{10581996260324}, \frac{7502876735617}{28562942440499}].$$

So the required d must be one of the denominates of the above covergents, but we do not know which one, so we just try

$$(3^{7502876735617})^3 \equiv 17387646817554 \not\equiv 3 \pmod{28562942440499}$$

$$(3^{7502876735617})^4 \equiv 7072755623312 \not\equiv 3 \pmod{28562942440499}$$

$$(3^{7502876735617})^{19} \equiv 11902526494611 \not\equiv 3 \pmod{28562942440499}$$

$$(3^{7502876735617})^{99} \equiv 5513494147015 \not\equiv 3 \pmod{28562942440499}$$

$$(3^{7502876735617})^{118} \equiv 8201089526821 \not\equiv 3 \pmod{28562942440499}$$

$$(3^{7502876735617})^{217} \equiv 8739051274402 \not\equiv 3 \pmod{28562942440499}$$

$$(3^{7502876735617})^{769} \equiv 3 \pmod{28562942440499}$$

So, $d = 769$. Clearly, for $N = 28562942440499$, Wiener's attack works well for

$$d < \frac{\sqrt[4]{N}}{3} \approx 770.6.$$

8.3 Extended Diophantine Attacks

Wiener's Diophantine attack on short RSA private exponent d applies to the case that

$$d < N^{0.25},$$

or more precisely,

$$d < \frac{1}{3}N^{0.25}.$$

About 10 years later, Boneh and Durfee [38] improved Wiener's method by showing that RSA is insecure for any

$$d < N^{0.292}.$$

Tha is,

$$\{e, N\} \xrightarrow[d \,<\, N^{0.292}]{\mathcal{P}} \{d\}. \tag{8.9}$$

For example, when $N = 160523347$, $d < 160523347^{0.292} \approx 248.89$, and when $N = 28562942440499$, $d < 28562942440499^{0.292} \approx 8493.651934$ are not secure. Their idea is as follows.

Let $d < N^{\delta}$. Notice first that

$$\begin{aligned} 1 &= ed + \frac{k\phi(N)}{2} \\ &= ed + k\left(\frac{N+1}{2} - \frac{p+q}{2}\right) \\ &= ed + k(x+y) \end{aligned}$$

Then finding d is equivalent to find the two small solutions x and y to the congruence

$$f(x, y) \equiv k(x+y) \equiv 1 \pmod{e}.$$

Letting

$$s = -\frac{p+q}{2}, \quad A = \frac{N+1}{2},$$

then

$$k(A+s) \equiv 1 \pmod{e}.$$

Let $e = N^{\alpha}$ for some α. So α is close to 1, as e is in the same order of the length of N. As d is small, so

$$|s| < 2\sqrt{N} = 2e^{1/(2\alpha)} \approx \sqrt{e},$$

and

$$|k| < \frac{2ed}{\phi(N)} \leq \frac{3ed}{N} < 3e^{1+\frac{\delta-1}{\alpha}} < e^{\delta}$$

By taking $\alpha \approx 1$ and ignoring small constants, we end up with the following *small inverse problem* [38]: finding integers k and s such that

$$k(A+s) \equiv 1 \pmod{e}, \tag{8.10}$$

where

$$|s| < e^{0.5} \quad \text{and} \quad |k| < e^{\delta}.$$

Given $\delta = 0.5$, one can list efficiently all solutions to the *small inverse problem*, thus RSA with $d < N^{\delta}$ is not secure, which is essentially Wiener's result [324]. What Boneh and Durfee [38] has shown is that whenever

$$\delta < 1 - \frac{1}{2} \approx 0.292$$

all small solutions to (8.10) can be efficiently calculated. This implies that for $\delta = 0.292$, RSA with $d < N^{\delta}$ is not secure. Boneh and Durfee admitted that this attack cannot be stated as a theorem, as they were unable to prove that it will always success, but they also could not find a counterexample that the attack failed. It is interesting to note that Steinfeld et al showed in [306] that a subexponential-time attack on RSA with private key $d > N^{0.25}$ cannot be obtained just by using Wiener's approach. So, the lattice-based attacks of [38] for $d < N^{0.292}$ with *heuristically* succeed in polynomial-time is currently an interesting open problem.

It is reasonable to conjecture that

Conjecture 8.3.1. Let $N = pq$ and $d < N^{0.5}$. Then given (e, N) Eve can efficiently compute d.

Of course, to prove or disprove this conjecture, some deep ideas and sophisticated techniques from mathematics will be needed.

Remarkable enough, Wiener's Diophantine attack on short RSA private exponent d can be naturally extended to other RSA-type cryptosystems. For example, Pinch [237] observed that Wiener's attack can be extended to cryptosystems based on groups whose order modulo p is closed to p such as the LUC cryptosystem [305] and the elliptic curve cryptosystems ECC (e.g., KMOV [180] and Demytko [98]). The encryption and decryption of a simplified version of LUC can be described as follows (other simplified version of LUC can be found in [205]):

[1] *Encryption:* Let M be the original message and write

$$\mu \equiv M + \sqrt{M^2 - 1}$$

as the plaintext. Then the message is encrypted by writing:

$$\mu^e \equiv C + \sqrt{C^2 - 1}$$

[2] *Decryption:* The decryption is then by using d in the following way:

$$\mu^{ed} \equiv C^d + \sqrt{(C^d)^2 - 1}$$

where d is one of the four values satisfying

$$ed_{\pm\pm} \equiv 1 \pmod{\operatorname{lcm}(p \pm 1, q \pm 1)}$$

according to the Jacobi symbols

$$\left(\frac{M^2-1}{p}\right) = \left(\frac{C^2-1}{p}\right)$$
$$\left(\frac{M^2-1}{q}\right) = \left(\frac{C^2-1}{q}\right)$$

Fixing the sign for the two Jacobi symbols and putting

$$ed = 1 + \frac{k}{h}(p \pm 1)(q \pm 1)$$

where

$$h = \text{lcm}(p \pm 1, q \pm 1)$$

then we have

$$\frac{e}{N} - \frac{k}{hd} = \frac{k}{hd}\left(\frac{1}{N} \pm \frac{1}{p} \pm \frac{1}{q}\right) + \frac{1}{dN}$$

so, $\frac{e}{N}$ will have $\frac{k}{hd}$ as its continued fraction expansion, same as that in Wiener's attack on RSA. Pinch also showed that Wiener's attack can be naturally extended to attack elliptic curve cryptosystems such as KMOV [180] and Demytko [98]. Interested readers are suggested to consult Pinch's original paper [237].

8.4 Small Private CRT-Exponent Attacks

To speed up the computation in RSA encryption and signature verification, public exponents e are often chosen to be small, for example, $e = 3, 17$ and 65537. However, the private exponent d should *never* be small. This makes the decryption and the signature generation slow. Luckily, there is a way to speed up the computation in d related exponentiations by using the Chinese Remainder Theorem (CRT).

Theorem 8.4.1. Let $N = pq$ with p and q prime numbers. Suppose

$$M_p \equiv C^d \pmod{p} \equiv C^{d \pmod{p-1}} \equiv C^{d_p} \pmod{p-1}, \tag{8.11}$$

and

$$M_q \equiv C^d \pmod{q} \equiv C^{d \pmod{q-1}} \equiv C^{d_q} \pmod{q-1}. \tag{8.12}$$

Then

$$M \equiv C^d \pmod{N} \tag{8.13}$$

can be computed by the following formula:

$$M \equiv (M_p \cdot q \cdot q^{-1} \bmod p + M_q \cdot p \cdot p^{-1} \bmod q) \pmod{N} \tag{8.14}$$

Note that the d_p, d_q defined by

$$d_p \equiv d \pmod{p-1} \tag{8.15}$$

$$d_q \equiv d \pmod{q-1} \tag{8.16}$$

in (8.11) and (8.12) are called the *RSA CRT-private exponents.*

Example 8.4.1. Let

$$\begin{aligned}
e &= 9007, \\
M &= 200805001301070903002315180419000118050019172105011309_ \\
&\quad 190800151919090618010705, \\
N &= 114381625757888867669235779976146612010218296721242362_ \\
&\quad 562561842935706935245733897830597123563958705058989075_ \\
&\quad 147599290026879543541. \\
C &= 968696137546220614771409222543558829057599911245743198_ \\
&\quad 746951209308162982251457083569314766228839896280133919_ \\
&\quad 90551829945157815154 \pmod{N}.
\end{aligned}$$

For decryption, since in any case, (p, q) and d are known to the authorized person who does the decryption:

$$\begin{aligned}
p &= 349052951084765094914784961990389813341776463849338 78_ \\
&\quad 43990820577, \\
q &= 327691329932667095499619881908344614131776429679929 42_ \\
&\quad 539798288533.
\end{aligned}$$

Thus, instead directly using Algorithm 1.3.5 to do the exponentiation C^d mod N with a large (d, N), we make a combined use of the Chinese Remainder Theorem (2.15) and Algorithm 1.3.5 on C^{d_p} mod p and C^{d_q} mod q with small (d_1, d_2, p, q), compared to (d, N), as follows:

$$\begin{aligned}
M_p &\equiv C^{d_p} \equiv C^{d \pmod{p-1}} \pmod{p} \\
&= 655836613815976230413021134466880014757055030407533996250_ \\
&= 695585 \\
M_q &\equiv C^{d_q} \equiv C^{d \pmod{q-1}} \pmod{q} \\
&= 272394108546908412756683639303179381441152113513972196544_ \\
&= 60909824
\end{aligned}$$

Thus,

$$M \equiv M_p \bmod p$$

$$M \equiv M_q \bmod q$$

Using CRT, we get:

$$
\begin{aligned}
M &\equiv (M_p \cdot q \cdot q^{-1} \bmod p + M_q \cdot p \cdot p^{-1} \bmod q) \pmod N \\
&\equiv 67213475823359647855158938441101049414616703225644246 5\_ \\
&\quad 02625084457504305918855175267000270823022321438065933 3\_ \\
&\quad 968103555797687889184\ + \\
&\quad 47168149934529219814076841535045562595601593495598316 86\_ \\
&\quad 49380595491056316420591415637149027601084886322323325 50\_ \\
&\quad 940853319809665062 \\
&\equiv 20080500130107090300231518041900011805001917210501130 9\_ \\
&\quad 1908001519190906180107 05 \pmod N
\end{aligned}
$$

So, we have done, and the plaintext is:

THE MAGIC WORDS ARE SQUEAMISH OSSIFRAGE.

Remark 8.4.1. Exponentiation by a combined use of CRT and Algorithm 2.3.1 on $C^{d_p} \bmod p$ and $C^{d_q} \bmod q$ is four time faster than the direct use of Algorithm 2.3.1 on $C^d \bmod N$. First note that computing both $M_p \equiv C^{d_p}$ and $M_q \equiv C^{d_q}$ require half as many multiplications as computing $M \equiv C^d$ directly. In addition, all intermediate values during the computation of $M_p \equiv C^{d_p}$ and $M_q \equiv C^{d_q}$ are only half as big as that would be in computing $M \equiv C^d$, as they are in the range $[1, p]$ rather than $[1, N]$. When quadratic time complexity is considered for multiplications (faster algorithm for multiplication exists, but for all practical purposes, quadratic time complexity is sufficient), multiplying two numbers in $\mathbb{Z}_p$ takes 1/4 the time as multiplying two numbers in $\mathbb{Z}_N$. Hence, computing M_p takes 1/8 the time of computing M directly, and computing M_q also takes 1/8 the time of computing M directly, so computing both M_p and M_q takes 1/4 of the time of computing M directly. Thus, CRT exponentiation is four time fast than direct exponentiation. Note that the final CRT exponentiation step

$$M \equiv (M_p \cdot q \cdot q^{-1} \bmod p + M_q \cdot p \cdot p^{-1} \bmod q) \pmod N$$

takes negligible time since it only needs to be done *once* and for all!

RSA decryption and signature verification with CRT are four time faster than standard RSA, because they are working modulo p and q rather than $N = pq$. However, there is a very efficient attack on RSA with CRT when d_p and d_q are small due to [157].

Theorem 8.4.2. (Efficient Attack on RSA-CRT for small d_p, d_q [157]). Let $N = pq$ be a β-bit RSA modulus with p and q primes of $\beta/2$ bits. Let $e < \phi(N)$, $d_p < p-1$ and $d_q < q-1$ be the public and private CRT-exponents, respecyively, satisfying

$$ed_p \equiv 1 \pmod{p-1}$$

and

$$ed_q \equiv 1 \pmod{q-1}.$$

Let also bitsize$(d_q) < \delta\beta$. Then N can be factorized in time polynomial in $\log N$, provided that

$$\delta < 0.073 - \epsilon.$$

The proof of the above theorem is based on a well-known heuristic assumption; interested readers are advised to consult the original paper [157]. What this theorem says is that there is a polynomial-time attack on RSA with private CRT-exponents

$$d_p, d_q < N^{0.073}.$$

In other words, N can be factorized in polynomial-time if $d_p, d_q < N^{0.073}$. That is,

Corollary 8.4.1.

$$\{e, N\} \xrightarrow[d_p, d_q \ < \ N^{0.073}]{\mathcal{P}} \{p, q\}. \tag{8.17}$$

8.5 Partial Private Key Exposure Attacks

Let (d, p, q) with $pq = N$ be the RSA private information. Suppose by some means say e.g., from side-channels, the cryptanalyst, Eve, is able to expose a fraction of the bits of d, or p, or q (from an information-theoretic point of view, this reduces the key entropy [321]), it is possible for Eve to find the rest of the bits of d, or p, or q. The first significant result in partial private key exposure attacks is due to Coppersmith in 1997 [78].

Suppose $N = pq$ and the $\beta/4$ most significant bits (Coppersmith called these bits as high-order bits) of p are known. Then by division the $\beta/4$ most significant bits of p is also known. Set

$$p = p_0 + x_0$$
$$q = q_0 + y_0$$

where p_0 and q_0 are known but p, q, x_0, y_0 are unknown. Define the bounds X, Y for x_0, y_0 as follows:

$$X = p_0 N^{-1/4} > |x_0|$$

$$Y = q_0 N^{-1/4} > |y_0|$$

Define the polynomial $p(x, y)$ as follows:

$$\begin{aligned} p(x,y) &= (p_0 + x)(q_0 + y) - N \\ &= q_0 x + p_0 y + xy + (p_0 q_0 - N). \end{aligned}$$

One integer solution of $p(x, y) = 0$ is given by (x_0, y_0), namely,

$$p(x_0, y_0) = pq - N.$$

Now we need a Lemma also due to Coppersmith:

Lemma 8.5.1. Let $p(x, y)$ be an irreducible polynomial in two variables over $\mathbb{Z}$, of maximum degree δ in x and y, respectively. Let X, Y be the bounds on the desired solution (x_0, y_0). Define $\tilde{p}(x, y) = p(xX, yY)$ and let W be the absolute value of the largest coefficient of $\tilde{p}$.

(1) If

$$XY < W^{2/(3\delta)-\epsilon} 2^{-148/3},$$

then in time polynomial in $(\log W, \delta, 1/\epsilon)$, all integer solutions (x_0, y_0) with $p(x_0, y_0) = 1$, $|x_0| < X$, and $|y_0 < Y|$ can be found.

(2) If

$$XY \leq W^{2/(3\delta)}$$

then in time polynomial in $(\log W, 2^\delta)$, all integer solutions (x_0, y_0) with $p(x_0, y_0) = 1$, $|x_0| < X$, and $|y_0 < Y|$ can be found.

we have

$$\begin{aligned} W &= \max_{ij}(|p_{ij}|X^i Y^j) \\ &= \max(|p_0 q_0 - N|, q_0 X, p_0 Y, XY) \\ &= N^{3/4} \end{aligned}$$

and

$$\begin{aligned} XY &= p_0 q_0 N^{-1/2} \\ &\approx N^{1/2} \\ &= W^{2/(3\delta)} \end{aligned}$$

which gives the following theorem:

Theorem 8.5.1 (Coppersmith).] Let $N = pq$ be a β bit RSA modulus such that the prime p is about $\beta/2$ bits. Then given the $\beta/4$ *most* significant bits of p, one can factor N in polynomial-time.

Similarly, if the $\beta/4$ least significant bits of p is given, one can also find factor N in polynomial-time:

Theorem 8.5.2. Let $N = pq$ be a β bit RSA modulus such that the prime p is about $\beta/2$ bits. Then given the $\beta/4$ *least* significant bits of p, one can factor N in polynomial-time.

Interested reader should consult [78] for the full justification of the above two theorems; these two theorems can in fact be conveniently combined into one single theorem as follows:

Theorem 8.5.3 (Coppersmith [78]). Let $N = pq$ be a β bit RSA modulus such that the prime p is about $\beta/2$ bits. Then given either the $\beta/4$ *least* significant bits of p, or the $\beta/4$ *most* significant bits of p, there is a polynomial-time algorithm to find the prime factorization of $N = pq$.

Remark 8.5.1. Rivest and Shamir [265] showed in 1986 that N can be efficiently factorized if the $\beta/3$ of the bits of p are given. Note also that Lehmer [184] in 1907 showed that if $|p - q| < 2N^{1/4}$ is given, N can be efficiently factorized. Thus, Coppersmith's result is significantly better than the previous results.

The next theorem, due to Boneh et al [39], shows that if the public exponent e is sufficiently small, then given a few of the bits of d, all of the d can be found in polynomial-time. As this attack requires e to be sufficiently small, one may also consider this attack as a short public exponent e attack.

Theorem 8.5.4 (BDF [39]). Let N be a β bit RSA modulus. Then given the $\beta/4$ least significant bits of d, one can find all of the d in time $\mathcal{O}(e \log e)$.

Proof. For $N = pq$ and $ed \equiv 1 \pmod{\phi(N)}$, there is an integer k such that

$$ed - k\phi(N) = ed - k(N - p - q + 1) = 1.$$

Since $d < \phi(N)$, then $0 < k \le e$. Multiply the equation by p, set $q = N/p$ and reduce the equation modulo $2^{\beta/4}$, we get:

$$p(ed) - kp(N - p + 1) + kN \equiv p \pmod{2^{\beta/4}}.$$

In this congruence, the cryptanalyst, Eve, knows (N, e) and the $\beta/4$ least significant bits of d, therefore, she knows the value of ed modulo $2^{\beta/4}$. Consequently, she obtains an equation in k and p. For each of the e possible values of k, Eve solves the quadratic equation in p, and obtaines a number of candidate values for p modulo $2^{\beta/4}$. For each of these candidate values of

p (although she is not sure which one is), she runs an algorithm of Theorem 8.5.3 to attempt to factor N. It can be shown that there are no more than $e \log_2 e$ candidate values for p modulo $2^{\beta/4}$. Hence, after at most $e \log_2 e$ attempts, N can be factored.

□

Remark 8.5.2. To defend against the partial private key exposure attack, keep the entirety of d secure, and try not to use small public exponent e, since if small e is used, the RSA system leaks *half* of the most significant bits of the corresponding private exponent d [37].

Example 8.5.1. Let $N = 1633 = 23 \cdot 71$ and $e = 23 = (11001100001)_2$ with 11 binary bits. Suppose Eve by some way knows 3 least significant bits of d, then by Theorem 8.5.4, she can find both d and the factorization of N. Now suppose that the three known bits are 011, which is 3 in decimal, and we let $d_0 = 011$. Then Eve does the following:

[1] As

$$ed_0 \equiv 1 + k(N - s + 1) \pmod{2^{n/4}}$$

where $s = p + q$, thus,

$$\begin{aligned}
&23 \cdot 3 \equiv 1 + k(1633 - s + 1) \pmod{2^3} \\
&\Longrightarrow 69 - 1 \equiv k(1633 - s + 1) \pmod{8} \\
&\Longrightarrow 4 \equiv k(1633 - s + 1) \pmod{8} \\
&\Longrightarrow s \equiv 6 \pmod{8}.
\end{aligned}$$

[2] Eve then solves the quadratic polynomial equation:

$$\begin{aligned}
&p^2 - sp + N \equiv 0 \pmod{2^{n/4}} \\
&\Longrightarrow p^2 - 6p + 1633 \equiv 0 \pmod{2^3} \\
&\Longrightarrow p^2 - 6p + 6 \equiv 7 \pmod{2^3} \\
&\Longrightarrow p = 7, \\
&\qquad p = 3.
\end{aligned}$$

[3] Let $p_0 \equiv 7 \equiv p \pmod{8}$ and $q_0 \equiv q \pmod{8}$. Then

$$\begin{aligned}
&p_0 q_0 \equiv 1633 \\
&\Longrightarrow 7q_0 \equiv 1 \pmod{8} \\
&\Longrightarrow q_0 \equiv 7 \pmod{8}.
\end{aligned}$$

[4] Create and solve the polynomial equation

$$\begin{aligned} & f(x, y) = 0 \\ \Longrightarrow & (rx + p_0)(ry + q_0) - N = 0 \\ \Longrightarrow & (8x + 7)(8y + 7) - 1633 = 0 \\ \Longrightarrow & x = 2, \\ & y = 8 \\ \Longrightarrow & p = 8 \cdot 2 + 7 = 23, \\ & q = 8 \cdot 8 + 7 = 71. \end{aligned}$$

[5] Find d

$$\begin{aligned} & ed - k\phi(N) = 1 \\ \Longrightarrow & 23d - (23 - 1)(71 - 1) = 1 \\ \Longrightarrow & 23d - 1540k = 1 \\ \Longrightarrow & d = 67. \end{aligned}$$

So, $d = 67$. Hence, any message using the given $(N, e) = (1633, 23$ as encryption key can be decrypted by using the computed private exponent $d = 67$.

8.6 Chapter Notes and Further Reading

The attacks directly related to the private exponent d are the most important attacks on RSA, since once d is found, any ciphertext using d for decryption can be broken. Much attention is of course given to find the d corresponding to the e. In this chapter, several attacks, such as the Diophantine attack, the lattice attack, and the CRT attack are discussed. These attacks all have the same objective. In quantum attacks, factoring attacks and discrete logarithm attacks, one essentially tries to find d using brute force approach. Whereas in the attacks discussed in this chapter one tries to find d using some special mathematical methods. These attacks work well for some special improper use and improper implementation of RSA. The Diophantine attack for small d was first studied by Winer in [324], and subsequently improved by Boneh and Durfee [38]. Lattice attacks are powerful methods for breaking RSA in some cases and have been studied by many authors, particularly Coppersmith [78]. In RSA encryption and signature verification, the public exponent e some times can be chosen to be small. However, the private exponent d should *never* be small. This means in RSA, the decryption and digital signature generation can be very slow, since d is large. There is a way to speed up the computation in d related exponentiations by using the Chinese Remainder Theorem (CRT), but there is a risk in using such a speeding method, since it may enable the enemy to discover the prime factors p and q of N if there is an unexpected error in the computation of the Chinese Remainder Theorem process. Thus, whenever possible, we should try to avoid the use of such CRT speeding-up techniques. For an enemy cryptanalyst, it may also be possible to discover d even by knowing a small portion of the bits of d. Thus, all bits of d should always be kept in a safe place.

There are many papers on the RSA short private exponent d attacks, with Winer's [324] being the first. Interested readers are advised to consult the following references for more information: Ernst et al [107], Hinek ([141] and [142]), Blöm and May [34], Boneh [37], Boheh and Venkatesan [40], Boneh and Durfee [38], Coron [83], Kim and Quisquater [167], Pfitzmann and Waidne [240], Stamp and Low [310], Steinfeld et al [306], and Yen at al [338]. Misarsky [214] proposed a multiplicative attack using the LLL algorithm on RSA signatures with redundancy.

9. Side-Channel Attacks

The development of cost-effective and efficient countermeasures to side-channel attacks is an ongoing research problem that is bing tackled by both cryptographers and engineers.

NEAL KOBLITZ AND ALFRED J. MENEZES
A Survey of Public-Key Cryptography [177]

9.1 Introduction

In the previous chapters, we have investigated two main types of mathematical and algorithmic attacks on RSA: direct (brute-force) attacks on the RSA modulus N including factoring, discrete logarithm and quantum attacks, and indirect attacks on RSA parameters $N, e, d, \phi(N), C$ including common modulus attacks, fixed-point attacks, and short exponent e and d attacks. A common feature of these attacks is that they are all directly on the RSA algorithm or more generally the RSA system, particularly the weakness of the algorithm and the system. There is, however, a large class of attacks, which are not directly on the RSA algorithm or the system itself. That is, they do not try to factor the modulus N and also do not exploit any inherent weakness issues of the RSA algorithm and system itself. Instead, they exploit specific implementation issues on a particular computing device, for example, a smartcard or a computer, that allow the secret information, i.e., the private exponent d to leak. These attacks are usually referred to as *side-channel attacks* or *implementation attacks*; they include but are not limited to *timing attacks* and *power attacks*. These attacks, although originally developed and applied for hardware security tokens such as smartcards, are applicable to general purpose hardware and software, as soon as the computing time is dependant on the implementation or execution. For example, Brumley and Boneh [58] developed some timing attacks on the general purpose web servers, OpenSSL server, since the OpenSSL's implementation of RSA is based on using the

Chinese Remainder Theorem and Montgomery reduction, which are vulnerable to the timing attacks. Furthermore, Bortz et al [42] even showed that by timing the web sites to take the response to HTTP requests, the client's private information can be leaked. More importantly, the side-channel attacks are practical and efficient. Thus, a secure network system must be timing attack resistant. In this chapter, we shall study various approaches to the side-channel attacks on RSA, particularly the time and power attacks.

9.2 Modular Exponentiation Revisited

The modular exponentiation:

$$x \mapsto x^k \bmod N$$

is the most fundamental operation in RSA encryption and decryption as well as RSA signature generation and verification. Now suppose given C, d, N, we want to do the RSA decryption:

$$M \equiv C^d \pmod{N}.$$

Let

$$d = \sum_{i=0}^{\beta} d_i 2^i = d_\beta d_{\beta-1} \cdots d_2 d_1 d_0$$

be the binary representation of d. Then the RSA decryption can be implemented by one of the following two algorithms.

The first algorithm performs the repeated squaring according to the binary bits of d beginning from the least significant bit to the most significant bit.

Algorithm 9.2.1. (Fast exponentiation by repeated squarings and multiplications beginning from the least significant bit)

> Input C and N
> Let $d = d_\beta d_{\beta-1} \cdots d_2 d_1 d_0$
> $M \leftarrow 1$
> for i from 0 to β do
>   If $d_i = 1$ then $M \leftarrow M \cdot C \pmod{N}$
> $C \leftarrow C^2 \pmod{N}$
> Print M

Or, equivalently and more conveniently,

Input d, C and N
$M \leftarrow 1$
While $d \geq 1$ do
  If $d \bmod 2 = 1$ then $M \leftarrow M \cdot C \pmod N$
  $C \leftarrow C^2 \pmod N$
  $d \leftarrow \lfloor d/2 \rfloor$
Print M

The second algorithm performs the repeated squaring according to the binary bits of d beginning from the most significant bit down to the least significant bit.

Algorithm 9.2.2. (Fast exponentiation by repeated squarings and multiplications beginning from the most significant bit)

Input C and N
Let $d = d_\beta d_{\beta-1} \cdots d_2 d_1 d_0$
$M \leftarrow 1$
for i from β down to 0 do
  $C \leftarrow C^2 \pmod N$
  If $d_i = 1$ then $M \leftarrow M \cdot C \pmod N$
Print M

Note that these two algorithms for modular exponentiations can be extended to the group operations on elliptic curves

$$Q \equiv kP \pmod N$$

where P and Q are points on elliptic curve $E : y^2 = x^3 + ax + b$ over $\mathbb{Z}_n$. Just the same as modular exponentiations in RSA cryptography, the elliptic curve group operations are the most fundamental operations in Elliptic Curve Cryptography (ECC).

Now let us move on to the next section on the discussion of implementation attacks on RSA where the modular exponentiation plays a significant role.

9.3 Timing Attacks

In 1995, Kocher [178], then an undergraduate student at Stanford University, surprised the cryptographic world by proposing a completely different new attack on RSA to get the private exponent d without factoring N. The attack is nothing to do with the RSA algorithm/system itself, rather, it has something to do with the hardware implementation of the RSA algorithm/system.

Kocher noticed that by measuring the *time* taken by the hardware (e.g., a smartcard or a computer) to perform the RSA decryption or signature that uses the modular exponentiation algorithm, it is possible to find the private exponent d without factoring N by using some knowledge of probability and statistics. The idea is as follows. The attacker first observes k, measuring the time T_i taken by the victim to compute each M_i for $i = 1, 2, \cdots, k$ as follows:

$$\begin{array}{ll} M_1 \equiv C_1^d \pmod N & \text{with time } T_1, \\ M_2 \equiv C_2^d \pmod N & \text{with time } T_2, \\ \vdots & \vdots \\ M_k \equiv C_k^d \pmod N & \text{with time } T_k, \end{array}$$

provided that the attacker knows (C_i, N) and the victim uses Algorithm 9.2.1 to implement the RSA decryption. The attack will allow somebody who knows the exponent $d_0, d_1, \cdots, d_{s-1}$ to find the bit d_s; to obtain the entire exponent d, start from d_0, repeat the attack until the entire exponent $d = d_0, d_1, \cdots, d_{s-1}, d_s, \cdots, d_\beta$ is known. Now, we start from d_0, the least significant bit of d. Since d is odd, we know that $d_0 = 1$. At this stage, we have:

$$d_0 = 1, \quad M \equiv C, \quad C \equiv C^2 \pmod N.$$

Next, we consider d_1. If $d_1 = 1$, then the victim will need to compute one modular multiplication and one modular squaring:

$$M \leftarrow M \cdot C \pmod N, \quad C \leftarrow C^2 \pmod N$$

So, at this stage, we have:

$$\begin{array}{lll} d_0 = 1, & M \equiv C, & C \equiv C^2 \pmod N \\ d_1 = 1, & M \equiv M \cdot C \equiv C^3 \pmod N, & C \equiv C^4 \pmod N. \end{array}$$

However, if $d_1 = 0$, then the victim will only need to compute one modular squaring without doing the modular multiplication:

$$C \leftarrow C^2 \pmod N$$

So, at this stage, we have:

$$\begin{array}{lll} d_0 = 1, & M \equiv C, & C \equiv C^2 \pmod N \\ d_1 = 0, & M \equiv C, & C \equiv C^4 \pmod N). \end{array}$$

Let t_i be the time it takes the hardware to compute $M_i \cdot C_i \equiv M_i \cdot M_i^2 \pmod N$. Of course, the t_i's different from each other, since the time to calculate $M_i \cdot M_i^2 \pmod N$ depends on the values of M_i, and also, the attacker does not yet know whether or not this multiplication actually occurs.

To give a formal analysis of the timing attack, we need some basic concepts from probability theory [268]. Suppose we have a random process (algorithm) that produces real numbers u as output where u is the amount of time it takes the computer to perform a calculation for a given input. If we record outputs $u_1, u_2, \cdots, u_k$ for a given input, then the *mean* (average, expected value) m is the average of these outputs:

$$m = \frac{1}{k}\sum_{i=1}^{k} u_i. \tag{9.1}$$

The *variance* of the $\{u_i\}$ is defined to be:

$$\mathbf{var}\left(\{u_i\}_{i=1}^k\right) = \frac{1}{k}\sum_{i=1}^{k}(u_i - m)^2, \tag{9.2}$$

and the *standard derivation* is the square root of the variance $\sqrt{\mathbf{var}(\{u_i\}_{i=1}^k)}$. The important fact we need in our analysis is that when the random process $\{u_i\}$ is independent of other random process $\{v_i\}$, the variance of the sum of their outputs is the sum of variances of the two processes:

$$\mathbf{var}\left(\{u_i\}_{i=1}^k\right) + \mathbf{var}\left(\{v_i\}_{i=1}^k\right) = \mathbf{var}\left(\{u_i\}_{i=1}^k + \{v_i\}_{i=1}^k\right). \tag{9.3}$$

The most interesting thing Kocher observed is that when $d_i = 1$, $\{t_i\}$ and $\{T_i\}$ are correlated; if for some i, t_i is larger than its expected value, T_i is also likely to be larger than its expected value. On the other hand, if $d_i = 0$, $\{t_i\}$ and $\{T_i\}$ behave as independent random variables. Let $t'_i = T_i - t_i$. The attacker computes $\mathbf{var}\left(\{T_i\}_{i=1}^k\right)$ and $\mathbf{var}\left(\{t'_i\}_{i=1}^k\right)$. Thus, if $d_i = 1$ then $\mathbf{var}\left(\{T_i\}_{i=1}^k\right) > \mathbf{var}\left(\{t'_i\}_{i=1}^k\right)$, which is so, because

$$\mathbf{var}\left(\{T_i\}_{i=1}^k\right) = \mathbf{var}\left(\{t_i\}_{i=1}^k\right) + \mathbf{var}\left(\{t'_i\}_{i=1}^k\right) > \mathbf{var}\left(\{t'_i\}_{i=1}^k\right). \tag{9.4}$$

However, if $d_i = 0$, then (9.4) will fail to hold. Hence, the attacker can determine d_1. Continuing the process this way, the attacker will eventually find $d_2, d_3, \cdots, d_\beta$, the entire exponent d.

Remark 9.3.1. When a short e is used, it is not necessary to use the time attack to find all the bits of d; only about $1/4$ of the bits of d are needed, since by using the $1/4$ bits of d, the *partial key exposure attack* discussed in Section 8.5, the attacker can discover all of the bits of d.

Kocher showed in [178] that the probability that a guess d_j for the first j bits of d is correct is proportional to

$$\mathrm{Prob}(d_i) \propto \prod_{i=1}^{k} F(T_i - t(C_i, d_j)) \tag{9.5}$$

where $C_1, C_2, \cdots, C_k$ are the ciphertext, $T_1, T_2, \cdots, T_k$ are the corresponding timing measurements for decrypting $C_1, C_2, \cdots, C_k$, $t(C_i, d_j)$ is the amount of time required for the first j iterations of the $M \equiv C_i^d \pmod{N}$ computation using the exponent d_j, and F is the expected probability distribution function of $T - t(C, d_j)$ over all C values and the correct d_j. Given a correct guess for d_{j-1}, there are two possible values for d_j. The probability that d_j is correct and d'_j is incorrect can be found as

$$\frac{\prod_{i=1}^{k} F(T_i - t(C_i, d_j))}{\prod_{i=1}^{k} F(T_i - t(C_i, d_j)) + \prod_{i=1}^{k} F(T_i - t(C_i, d'_j))} \tag{9.6}$$

A possible defense to the timing attack, suggested by Rivest [163], is as follows:

Algorithm 9.3.1 (Anti-Timing Attack). RSA Blinding.

[1] Generate a secret random number $r \in \mathbb{Z}_N^*$.
[2] Compute $C' \equiv Cr^e \pmod{N}$.
[3] Compute $M' \equiv (C')^d \pmod{N}$.
[4] Compute $M \equiv M'r^{-1} \pmod{N}$.

Theorem 9.3.1 (Correctness of Theorem 9.3.1). $M \equiv M'r^{-1} \pmod{N}$.

Proof. Theorem 9.3.1 is correct since

$$\begin{aligned} M'r^{-1} &\equiv (C')^d r^{-1} \\ &\equiv (Cr^e)^d r^{-1} \\ &\equiv C^d r^{ed} r^{-1} \\ &\equiv M \pmod{N}. \end{aligned}$$

□

9.4 Time Attacks on RSA in OpenSSL

SSL (Secure Sockets Layer) is one of the most popular systems securing the web. For example, when you see your web browser's application protocol changing from http to https, then all the cryptographic operations that comprise SSL are operated on your behalf to enhance the security of your data transfers between you and the server. OpenSSL, where RSA is embedded in, is a full-strength, general purpose cryptography library for SSL (version 2 and version 3) and TLS (version 1). To perform the decryption

$$M \equiv C^d \pmod{N},$$

in OpenSSL, it uses the Chinese Remainder Theorem (CRT) for fast computation. With CRT, $M \equiv C^d \pmod{N}$ is computed in two steps:

[1] Compute

$$\left.\begin{aligned} M_1 &\equiv C^{d_1} \pmod{p}, \\ M_2 &\equiv C^{d_2} \pmod{q}. \end{aligned}\right\} \tag{9.7}$$

[2] Use CRT to combine M_1 and M_2 to get M.

Although RSA with CRT is not vulnerable to Kocher's original time attack, time attacks on the exponentiation can be used to expose the factors of N, since RSA with CRT uses the factors p and q of N in its fast exponentiation as described in (9.7). Note that OpenSSL uses an optimization technique, called *sliding windows*, for the square and multiply operation. That is, when the sliding windows technique is used, a block of bit (in fact, 5 bits) of d, rather than only one bit of d in the conventional square and multiply technique, are processed at each of the iterations. Note also that the sliding windows exponentiation technique performs a modular multiplication at each step, using the Montgomery reduction technique, which transforms a reduction modulo e.g., p into a reduction modulo some powers of 2, denoted by R, since a reduction modulo some powers of 2 is faster than a reduction modulo q, as many arithmetic operations can be directly implemented in hardware, but of course, before using Montgomery reduction all variables must be written in *Montgomery form*, which is just $xR \bmod p$ for x. To perform, e.g., $c = ab$ in Montgomery form, just do:

$$\begin{aligned} aR \cdot bR &= cR^2, \\ cR^2 \cdot R^{-1} &\equiv cR \pmod{p}, \\ cR \cdot R^{-1} &\equiv c \pmod{p}. \end{aligned}$$

The last step above is just get back the stand form from the Montgomery form. More information on Montgomery reduction can be found in [219]. In what follows, we give a brief introduction to the Brumley-Boneh time attack on RSA with CRT in OpenSSL (see [58] and [310] for more detailed information).

Algorithm 9.4.1 (Time Attack on RSA with CRT in OpenSSL). Let $N = pq$ be the RSA modulus. This algorithm tries to find the bit information of the prime factor, say, e.g., p of N, since once p is found, d can be easily calculated by

$$d \equiv 1/e \pmod{(p-1)(q-1)}.$$

Let p have β bits, i.e.,

$$p = (p_0, p_1, p_2, \cdots, p_{\beta-1})_2.$$

Clearly, it is not necessary to use the time attack to find all the bits of p; only about $\beta/2$ of the bits of p are needed, since by Theorem 8.5.1, using these $\beta/2$ bits of p, N can be efficiently factorized.

[1] Let the bits $p_1, p_2, \cdots, p_{i-1}$ of p have been determined. Let P_1 represents the most significant bits (high bits) of p, respectively, with the remaining bits set to 0. That is,

$$P_1 = (p_0, p_1, \cdots, p_{i-1}, 0, 0, \cdots, 0).$$

Let also the ith bit of p set to 1. That is,

$$P_2 = (p_0, p_1, \cdots, p_{i-1}, 1, 0, \cdots, 0).$$

If the *unknown* bit p_i is 1, then

$$P_1 < P_2 \leq p,$$

otherwise,

$$P_1 \leq p < P_2.$$

[2] Let $T(P_1)$ and $T(P_2)$ be the computing time for P_1 and P_2, respectively. Let also

$$\Delta = |T(P_1) - T(P_2)|.$$

[3] If $P_1 < p < P_2$, then Δ will be *large*, and hence the bit p_i of p will be 0. If $P_1 < P_2 < p$, then Δ will be *small*, and hence the bit p_i of p will be 1.

[4] Repeat the process in order to find successively the

$$p_{i+1}, p_{i+2}, p_{i+3}, \cdots$$

until half of the bits of p are found, then p and q can be computed by Theorem 8.5.1.

There are of course several possible defenses for such a time attack on the RSA with CRT in OpenSSL; one of them is the blinding technique discussed in the previous section. Thus, OpenSSL should have, as it has already been implemented, blinding as its default operation to resist such time attacks.

Noted that Bortz, Boneh and Nangy [42] recently developed two different types of time attacks for exposing private information by timing the web applications. Surprisingly, their code (to use an invisible image and JavaScript to take several timing samples of the same or different pages sequentially) is extremely simple:

```
<html><body><img id="test" style="display: none">
<script>
   var test = document.getElementById('test');
   var start = new date();
   test.onerror = function() {
     var end = new date();
     alert("Total time: " + (end - start));
   }
   test.src = "http://www.example.com/page.html";
</script>
</body></html>
```

Interestingly, as the authors pointed out, these side-channel attacks are often ignored by web developers. Hence, future web applications should resist such attacks.

9.5 Power (Analysis) Attacks

A similar attack, called the *power (analysis) attack*, is by careful measuring the power consumption of a hardware device, say, a smartcard, during the computation of the RSA decryption or digital signature generation, in order to deduce the bit information of d. It works since during multiplication the computer's power consumption is necessarily higher than that for squaring. Hence, if the cryptanalyst measures the length of these high consumption episodes she can easily deduce the bits of d, just the same as measuring the time in the time attack. In what follows, we give an example of a power attack on the RSA signature scheme [176], in the same setting as the time attack in the previous section. Suppose the smartcard generates the signature as follows:

Input d, H and N
$S \leftarrow 1$
While $d \geq 1$ do
  If $d_i = 1$ t
    then $S \leftarrow S \cdot H \bmod N$
  $H \leftarrow H^2 \bmod N$

$d \leftarrow \lfloor d/2 \rfloor$
print S

As the power consumed by a modular squaring is less than that by a modular multiplication, by the careful measurement of the power consumption on the cryptographic operations in the algorithm, the sequence of multiplication and squaring can be revealed and hence, the binary string of d can be discovered (i.e., d can be found). Again, this attack can be thwarted by inserting some dummy operations so that at each iteration of the loop there are one modular squaring and one modular multiplication, as follows:

Input d, H and N
$S_0 \leftarrow 1$
While $d \geq 1$ do
for i from 0 to β do
    $S_1 \leftarrow S_0 \cdot H \bmod N$
    $S_0 \leftarrow S_{d_i}$
    $H \leftarrow H^2 \bmod N$
    $d \leftarrow \lfloor d/2 \rfloor$
Print S_0

Remark 9.5.1. This anti-power attack process decreases the efficiency of the signature generation. Moreover, it may still allow some other side-channel attacks. Old attacks are thwarted, but new attacks may still come. The attack and anti-attack are endless processes.

Remark 9.5.2. The idea of power (analysis) attacks can be generalized to, for example, *electromagnetic (analysis) attacks* and *template attacks* [8].

9.6 Random Fault Attacks

As we discussed in Section 8.4 of the previous chapter, RSA decryption and signature verification with CRT are four time faster than standard RSA, which is specifically and helpful and useful for smartcards and systems-on-chip technologies. But unfortunately, it is susceptible to software and hardware errors and an attack, in fact, a very simple attack, enables the cryptanalyst to factor N, and hence break the system completely. The idea of the attack is based on obtaining two signatures of the same message: one is correct, the other in incorrect. Suppose the smartcard first computes the hashing function $h(M)$ of the message M to produce the document H for signature:

$$H = h(M)$$

Then the card computes

$$S_p \equiv H^{d_p} \equiv H^{d \bmod p-1} \pmod{p} \tag{9.8}$$

$$S_q \equiv H^{d_q} \equiv H^{d \bmod q-1} \pmod{q}. \tag{9.9}$$

The final signature is computed by using CRT as follows:

$$S \equiv (S_p \cdot q \cdot q^{-1} \bmod p + S_q \cdot q \cdot q^{-1} \bmod q) \pmod{pq}. \tag{9.10}$$

Now suppose that there is an error in the computation of either S_p or S_q so that one of the values for S_p and S_q is wrong; the wrong values are denoted by $\tilde{S}_p$ or $\tilde{S}_q$. In any one of the two wrong cases, the wrong signature $\tilde{S}$ (instead of the correct S) is obtained. Once the cryptanalyst, say Eve, receives the $\tilde{S}$, she knows it is wrong since

$$\tilde{S} \not\equiv H \pmod{N}.$$

If S_p is wrong, then

$$\tilde{S}^e \not\equiv H \pmod{p} \tag{9.11}$$

$$\tilde{S}^e \equiv H \pmod{q} \tag{9.12}$$

therefore, by computing

$$q = \gcd(S^e - H, N) \tag{9.13}$$

the factor q of N was found. However, if S_q is wrong, then

$$\tilde{S}^e \equiv H \pmod{p} \tag{9.14}$$

$$\tilde{S}^e \not\equiv H \pmod{q} \tag{9.15}$$

therefore, just by computing

$$p = \gcd(S^e - H, N) \tag{9.16}$$

Eve finds the factor p of N. In what follows, we present a complete example of this attack.

Example 9.6.1. Let

$$
\begin{aligned}
e &= 9007, \\
H &= 200805001301070903002315180419000118050019172105011309\_ \\
&\quad 190800151919090618010705, \\
N &= 114381625757888867669235779976146612010218296721242362\_ \\
&\quad 562561842935706935245733897830597123563958705058989075\_ \\
&\quad 147599290026879543541. \\
p &= 3490529510847650949147849619903898133417764638493387\_ \\
&\quad 843990820577, \\
q &= 32769132993266709549961988190834461413177642967992942\_ \\
&\quad 539798288533, \\
d &= 106698614368578024442868771328920154780709906633937862\_ \\
&\quad 801226224496631063125911774470873340168597462306553968\_ \\
&\quad 544513277109053606095.
\end{aligned}
$$

Then

$$
\begin{aligned}
d_p &\equiv d \pmod{p-1} \\
&= 72856616857928097972665230214492378048466720554763729\_ \\
&\quad 8401547983 \\
d_q &\equiv d \pmod{q-1} \\
&= 31463024550435273030761771275045678061636533405929\_ \\
&\quad 051524833529391 \\
S_p &\equiv H^{d_p} \pmod{p} \\
&= 62669140930451146914695358216820729474704322050636\_ \\
&\quad 3221769613673 \\
S_q &\equiv H^{d_q} \pmod{q} \\
&= 23960980522391473260221131686329987188383969752491\_ \\
&= 892298428767968
\end{aligned}
$$

$$
\begin{aligned}
S_1 &\equiv (S_p \cdot q \cdot q^{-1} \bmod p) \pmod N \\
&\quad 883708025919394818509037855308020872961492832980459\_ \\
&\quad 614536119803759634744064181105100524778790714781138\_ \\
&\quad 4234300642369762485873012 6 \\
S_2 &\equiv (S_q \cdot p \cdot p^{-1} \bmod q) \pmod N \\
&\quad 217131130945912638991822885165014178473013148722592\_ \\
&\quad 894082951781397918073776638921361069039244477548457\_ \\
&\quad 2670705726878130278709258 2 \\
S &\equiv (S_1 + S_2) \pmod N \\
&\quad 110083915686530745750086074047303505143450598170305\_ \\
&\quad 250861907158515755281784082002646159381803519232959\_ \\
&\quad 5690500636924789276458227 08
\end{aligned}
$$

Once Eve received S, she know this signature is valid, since

$$
\begin{aligned}
S^e \pmod N &= H \\
&= 20080500130107090300231518041900011805001 9\_ \\
&\quad 172105011309190800151919090618010705.
\end{aligned}
$$

There is nothing Eve can do in this correct case. Now suppose that an error was introduced in the computation of S_p (S_q is still correct), causing $\tilde{S}_p = S_p + 2$:

$$
\begin{aligned}
S_p &= 626691409304511469146953582168207294747043220506363\_ \\
&\quad 22176961367\underline{5}
\end{aligned}
$$

such that

$$
\begin{aligned}
S_1 &\equiv (S_p \cdot q \cdot q^{-1} \bmod p) \pmod N \\
&\quad 455346113894307531087044864902191870931730603820281 9\_ \\
&\quad 428606011015945402988817864832367300730962942301244 9\_ \\
&\quad 350713341438023120501425 \\
S &\equiv (S_1 + S_2) \pmod N \\
&\quad 672477244840220170078867750067206049404743752542874836\_ \\
&\quad 943552882992458372658425404597799112340771778581760577\_ \\
&\quad 70610219325907594007
\end{aligned}
$$

This S is wrong, as Eve is readily to verify that

$$\begin{array}{rcl} S^e & \equiv & 96207603032468462723468660688826427395886111803586006\text{\_} \\ & & 57331991308787962552685185688083316222499132979809394 7\text{\_} \\ & & 3742509472164332131 5 \ (\text{mod } N) \\ & \neq & H. \end{array}$$

Thus, just by computing

$$\begin{array}{rcl} \gcd(S^e - H, N) & = & 32769132993266709549961988190834461413177642\text{\_} \\ & & 9679929425397982885 33 \\ & = & q. \end{array}$$

Eve finds the factor q of N. Alternatively, suppose that an error was introduced in the computation of S_q (S_p is still correct), causing $\tilde{S}_q = S_q + 2$:

$$\begin{array}{rcl} S_q & \equiv & 23960980522391473260221131686329987188383969752491 8\text{\_} \\ & & 922984287679\underline{70} \end{array}$$

such that

$$\begin{array}{rcl} S_2 & \equiv & (S_q \cdot p \cdot p^{-1} \bmod q) \ (\text{mod } N) \\ & = & 64549304297099992641381587557084318050277537788277 05\text{\_} \\ & & 65758470483563012518959033543224863744938898099471 19\text{\_} \\ & & 69935035104090452532128 5 \\ S & \equiv & (S_1 + S_2) \ (\text{mod } N) \\ & & 38538481131150606823049593111739793336208524365080655 4\text{\_} \\ & & 66897185796557791056587567001941728809002583001972967 2\text{\_} \\ & & 09175448502504507870. \end{array}$$

This S is wrong, as Eve is readily to verify that

$$\begin{array}{rcl} S^e & \equiv & 79282477674652789150217075014902535840218914449104029 7\text{\_} \\ & & 87978490721410501661318967919737342869474235857923565 1\text{\_} \\ & & 5808251175817416555 7 \ (\text{mod } N) \\ & \neq & H. \end{array}$$

Thus, just by computing

$$\begin{array}{rcl} \gcd(S^e - H, N) & = & 34905295108476509491478496199038981334177646\text{\_} \\ & & 3849338784399082057 7 \\ & = & p \end{array}$$

Thus, Eve finds the factor p of N and hence all the RSA trap-door one-way information such as $q, \phi(N)$ and d.

Remark 9.6.1. This attack is useful when Eve has the full knowledge of H, the document to be signed. However, if a random padding procedure was used before the signature, the attack can be defeated. Of course, if the signer can have a check on the correctness of S before before sending the signature out, the attack can also be defeated; this correctness checking is specifically useful and helpful when using RSA with CRT.

Remark 9.6.2. This attack for small private CRT-exponent is usually referred to as *random fault attack*, this type of attacks also belongs to *side-channel attacks*, because it is based on a miscalculation, which is, in turn, based on either a hardware or software error, known as a *glitch*, which is not directly related to the cryptographic algorithm or system itself.

A similar *random fault attach* is the Bleichenbacher's attack [32] on *PKCS#1*; it is a *chosen-ciphertext attack*. PKCS stands for Public-Key Cryptography Standard and is the RSA signature and encryption standard developed by RSA laboratories; a newer version of PKCS#1 is PKCS#1 v2.1 [271]. Let $N = pq$ be a β bit RSA modulus, and M be a γ bit message with $\beta > \gamma$. Prior to encryption, M is padded in the following way:

02	Random Pad	00	Plaintext Message M

where 02 is the 16 bit initial block indicating that a random pad has been added to the plaintext M. The results message is β bit long and is used for RSA encryption. Now suppose that Bob sends a PKCS#1 ciphertext to Alice, Alice decrypts it, checks the initial block, and throws off the random pad; if the initial block is not presented, Alice sends back an error message to Bob saying that "invalid ciphertext". Bleichenbacher observed that this error message can lead Eve, the cryptanalyst, to decrypt cyphertexts of her choice. The idea is as follows.

[1] Bob sends a PKCS#1 ciphertext C to Alice. Eve intercepts a ciphertext C.

[2] Eve Chooses a random number $r \in \mathbb{Z}_N^*$, computes

$$C' \equiv rC \pmod{N},$$

and sends C' to Alice.

[3] Alice receives C' and tries to decrypt it.

[4] If Alice does not send an error message back to Bob, Eve learns that the most significant 16 bits of C' are equal to 02.

[5] By accessing a *decryption oracle* that tests for Eve whether or not the most significant 16 bits of rC' are equal to 02 for any r of his choice, Eve will be able to decrypt C [32].

Two new attacks to PKCS#1 also exists, due to Coron et al [84], that do not make use of a *decryption oracle*, i.e., it is a *chosen-plaintext attack* against

PKCS#1. The first attack applies to small public exponents whereas the second applies for arbitrary public exponents. Interested readers are suggested to consult [84] for more information.

9.7 Chapter Notes and Further Reading

Time attacks were first proposed by Kocher [178] in 1995, then an undergraduate student at Stanford University; he conjectured that his technique can be generalized to attach cryptosystems where the modular exponentiation was computed using more efficient means. Of course, this is not true. Kocher's idea was made more practical by Dhem et al [95] and particularly by Schindler [279] to a more optimized fashion of the modular exponentiation, i.e., Montgomery's reduction. It had been pushed even further by Brumley and Boneh [58], who developed a successful time attack against the highly optimized RSA with CRT implementation in OpenSSL, and by Bortz et al [42], who showed that the time web site take to respond to HTTP requests can leak private information.

Time attacks belong to a wide range of class of attacks, called *side-channel attacks*, including *power attacks*, and the attacks based on *electromagnetic radiation*. Normally, all these attacks require some sort of physical access to machine/hardware, or some special equipment so that the operations on that device can be detected and measured. Interested readers are suggested to consult the following references and the references wherein for more information: Bortz et al [42], Boscher et al [43], Brier et al [56], Brumley and Boneh [58], Dhem et al [95], Fouque et al [113], Hachez and Quisquater [134], Kaliski [163], Klma and Rosa [168], Kocher [178], Kuhn [181], Mangard et al [195], Mayer-Sommer [198], Messerges et al [209], Naccache and Tunstall [226], Neuenschwander [230], Sakai and Sakurai [272], Schindler [279], Schindler and Walter [280], Schindley [275], Stamp and Low [310], and Walter [322].

Recently studies on side-channel attacks lead to a new area, *Micro-Architectural Cryptanalysis (MAC)*, which studies the effects of common processor components on cryptosystem security, as the microprocessor components generate easily observable execution (e.g., the *footprints* or *instruction paths*). So, it is very necessary to have some good methods to protect RSA against such attacks. Interested readers are suggested to consult [1], [2], [3], and the references therein.

10. The Road Ahead

Cryptography is a never-ending struggle between code makers and code breakers.

ADI SHAMIR
The 2002 Turing Award Recipient

10.1 Introduction

To date, RSA is still the most popular and secure public-key cryptographic system. However, it will become insecure if a practical quantum computer can be built, or some other advances in integer factorization and cryptanalysis. As Bill Gates pointed out in [117]:

> *We have to ensure that if any particular encryption technique proves fallible, there is a way to make an immediate transition to an alternative technique.*

So, if one day RSA is proved to be insecure, we have to shift RSA to some other cryptosystems. Fortunately, there are many alternative cryptosystems to replace RSA:

(0) Elliptic Curve cryptosystems (ECC), such as the Koblitz system [170],

(1) Coding-Based Cryptosystems (CBC), such as the McEliece system [203],

(2) Lattice-Based Cryptosystems (LBC), such as the Cai-Cusick system [62] and NTRU [144],

(3) Quantum Cryptography [26],

to name just a few. Note that we regard ECC as alternative 0 to RSA, since in theory, the levels of the difficulty for breaking ECC and RSA are almost the same. For example, as we have already discussed in the book, the quantum attacks are applicable to both RSA and ECC, and Wiener's attack for RSA with short e can also be extended to ECC such as KMOV. However, from practical point of view, for the same level of security, the length for ECC keys are significantly shorter than that for RSA keys. This makes ECC particularly useful in wireless communications and smartcards. The other three types of alternatives can be used as an immediate replacement to RSA whenever RSA becomes insecure, for example, when a practical quantum computer becomes available for use. In this chapter, we shall briefly discuss some of these alternatives to RSA.

10.2 Elliptic Curve-Based Cryptography

In Section 2.5, we discussed some elliptic curve analogues of RSA, whose security relies on the intractability of the Integer Factorization Problem (IFP). It should be noted that the security of the general Elliptic Curve Cryptography (ECC) does not rely on the intractability of IFP, but on the intractability of the Elliptic Curve Discrete Logarithm Problem (ECDLP), which is a generalization of the Discrete Logarithm Problem (DLP) on elliptic curve group $E/\mathbb{F}_q$ (over a large finite field), rather then on a multiplicative group $\mathbb{Z}_N^*$. The main feature of ECC is that for the same level of security, the key length for ECC is shorter than that for other types of public-key cryptosystems such as RSA. In what follows, we shall introduce an ECC system, due to Menezes and Vanstone [206].

[1] **Key generation:** Alice and Bob publicly choose an elliptic curve E over $\mathbb{F}_p$ with $p > 3$ is prime and a random *base* point $P \in E(\mathbb{F}_p)$ such that P generates a large subgroup H of $E(\mathbb{F}_p)$, preferably of the same size as that of $E(\mathbb{F}_p)$ itself. Assume that randomly chosen $k \in \mathbb{Z}_{|H|}$ and $a \in \mathbb{N}$ are secret.

[2] **Encryption:** Suppose now Alice wants to sent message

$$m = (m_1, m_2) \in \mathbb{Z}_p^* \times \mathbb{Z}_p^* \tag{10.1}$$

to Bob, then she does the following:

[2-1] $\beta = aP$, where P and β are public.

[2-2) $(y_1, y_2) = k\beta$

[2-3] $c_0 = kP$.

[2-4] $c_j \equiv y_j m_j \pmod{p}$ for $j = 1, 2$.

[2-5] Alice sends the encrypted message c of m to Bob:

$$c = (c_0, c_1, c_2). \tag{10.2}$$

[3] **Decryption:** Upon receiving Alice's encrypted message c, Bob calculates the following to recover m:

[3-1] $ac_0 = (y_1, y_2)$.

[3-2] $m = \left(c_1 y_1^{-1} \pmod p,\ c_2 y_2^{-1} \pmod p\right)$.

Example 10.2.1. The following is a nice example of Menezes-Vanstone cryptosystem, taken from [216].

[1] **Key generation:** Let E be the elliptic curve given by $y^2 = x^3 + 4x + 4$ over $\mathbb{F}_{13}$, and $P = (1, 3)$ be a point on E. Choose $E(\mathbb{F}_{13}) = H$ which is cyclic of order 15, generated by P. Let also the private keys $k = 5$ and $a = 2$, and the plaintext $m = (12, 7) = (m_1, m_2)$.

[2] **Encryption:** Alice computes:

$$\beta = aP = 2(1,3) = (12,8)$$
$$(y_1, y_2) = k\beta = 5(12,8) = (10,11)$$
$$c_0 = kP = 5(1,3) = (10,2)$$
$$c_1 \equiv y_1 m_1 \equiv 10 \cdot 2 \equiv 3 \pmod{13}$$
$$c_2 \equiv y_2 m_2 \equiv 11 \cdot 7 \equiv 12 \pmod{13}.$$

Then Alice sends

$$c = (c_0, c_1, c_2) = ((10,2), 3, 12)$$

to Bob.

[3] **Decryption:** Upon receiving Alice's message, Bob computes:

$$ac_0 = 2(10,2) = (10,11) = (y_1, y_2)$$
$$m_1 \equiv c_1 y_1^{-1} \equiv 12 \pmod{13}$$
$$m_2 \equiv c_2 y_2^{-1} \equiv 7 \pmod{13}.$$

Thus, Bob recoveres the message $m = (12, 7)$.

Compared to RSA, ECC requires shorter secret-key than RSA for the same level of security (see Table 10.1), but of course ECC requires longer secret-key than AES, since AES is a secret-key cryptosystem.

10.3 Coding-Based Cryptography

In this section, we introduce the most famous code-based cryptosystem, the McEliece system, invented by McEliece in 1978 [203]. One of the most important features of the McEliece system is that it has resisted cryptanalysis to

AES Key Size (bits)	ECC Key Size (bits)	RSA Key Size (bits)
	163	1024
	224	2048
128	256	3072
192	384	7680
256	512	15360

Table 10.1. Key Size Comparison among ECC, RSA and AES

date; it is even quantum computer resisted. The idea of the McEliece system is based on coding theory and its security is based on the fact that decoding an arbitrary linear code is $\mathcal{NP}$-complete.

Algorithm 10.3.1 (McEliece's Code-Based Cryptography). Suppose Bob wishes to send an encrypted message to Alice, using Alice's public-key. Alice generates her public-key and the corresponding private key. Bob uses her public-key to encrypt his message and sends it to Alice, Alice uses her own private-key to decrypt Bob's message.

[1] **Key Generation:** Alice performs:

[1-1]Choose integers k, n, t as common system parameters.

[1-2]Choose a $k \times n$ generator matrix G for a binary (n, k)-linear code which can correct t errors and for which an efficient decoding algorithm exists.

[1-3]Select a random $k \times k$ binary non-singular matrix S.

[1-4]Select a random $k \times k$ permutation matrix P.

[1-5]Compute the $k \times n$ matrix $\widehat{G} = SGP$.

[1-6]Now $(\widehat{G}, t)$ is Alice's public-key whereas (S, G, P) is Alice's private-key.

[2] **Encryption:** Bob uses Alice's public-key to encrypt his message to Alice. Bob performs:

[2-1]Obtain Alice's authentic public key $(\widehat{G}, t)$.

[2-2]Represent the message in binary string m of length k.

[2-3]Choose a random binary error vector z of length n having at most t 1's.

[2-4]Compute the binary vector $c = m\widehat{G} + z$.

[2-5]Send the ciphertext c to Alice.

[3] **Decryption:** Alice receives Bob's message m and uses her private-key to recover c from m. Alice does performs:

[3-1] Compute $\widehat{c} = cP^{-1}$, where P^{-1} is the inverse of the matrix P.

[3-2] Use the decoding algorithm for the code generated by G to decode $\widehat{c}$ to $\widehat{m}$.

[3-3] Compute $m = \widehat{m}\widehat{S}^{-1}$. This m is thus the original plaintext.

Theorem 10.3.1 (Correctness of McEliece's Cryptosystem). In McEliece's Cryptosystem, m can be correctly recovered from c.

Proof. Since

$$\begin{aligned}\widehat{c} &= cP^{-1} \\ &= (m\widehat{G} + z)P^{-1} \\ &= (mSGP + z)P^{-1} \\ &= (mS)G + zP^{-1}, \quad (zP^{-1} \text{ is a vector with at most } t\ 1's)\end{aligned}$$

the decoding algorithm for the code generated by G corrects $\widehat{c}$ to $\widehat{m} = mS$. Now applying S^{-1} to $\widehat{m}$, we get $mSS^{-1} = m$, the required original plaintext.

□

Remark 10.3.1. The security of McEliece's cryptosystem is based on error-correcting codes, particularly the Goppa [194]; if the Goppa code is replaced by other error-correcting codes, the security will be severely weakened. The McEliece's cryptosystem has two main drawbacks:

(1) the public-key is very large, and

(2) there is a message expansion by a factor of n/k.

It is suggested that the values for the system parameters should be $n = 1024$, $t = 50$, and $k \geq 644$. Thus for these recommended values of system parameters, the public-key has about 2^{19} bits, and the message expansion is about 1.6. For these reasons, McEliece's cryptosystem receives little attention in practice. However, as McEliece's cryptosystem is the first probabilistic encryption and more importantly, it has resisted all cryptanalysis including quantum cryptanalysis, it may be a good candidate to replace RSA in the post-quantum cryptography age.

10.4 Lattice-Based Cryptography

Cryptography based on ring properties and particularly lattice reduction is another promising direction for post-quantum cryptography, as lattice reduction is a reasonably well-studied hard problem that is currently not known to be solved in polynomial-time, or even subexponential-time on a quantum computer. There are many types of cryptographic systems based on lattice reduction. In this section, we give a brief account of one if the lattice based cryptographic systems, the NTRU encryption scheme. NTRU is rumored to stand for Nth-degree TRUncated polynomial ring, or Number Theorists eRe Us. It is a rather young cryptosystem, developed by Hoffstein, Pipher and Silverman [143] in 1995. We give a brief introduction to NTRU, for more information can be found in [143] and [144].

Algorithm 10.4.1 (NTRU Encryption Scheme). The NTRU encryption scheme works as follows.

[1] **Key Generation:**

[1-1] Randomly generate polynomials f and g in D_f and D_g, respectively, each of the form:

$$a(x) = a_0 + a_1x + a_2x^2 + \cdots + a_{N-2}x^{N-2} + a_{N-1}x^{N-1}.$$

[1-2] Invert f in $\mathcal{R}_p$ to obtain f_p, and check that g is invertible in f_q.

[1-3] The public-key is $h \equiv p \cdot g \cdot f_q \pmod q$. The private-key is the pair (f, f_p).

[2] **Encryption:**

[2-1] Randomly select a small polynomials r in D_r.

[2-2] Compute the ciphertext $c \equiv r \cdot h + m \pmod q$.

[3] **Decryption:**

[3-1] Compute $a = \text{center}(f \cdot c)$,

[3-2] Recover m from c by computing $m \equiv f_p \cdot a \pmod q$. This is true since

$$a \equiv p \cdot r \cdot \equiv +f \cdot m \pmod q.$$

In Table 10.2, we present some information comparing NTRU to RSA and McEliece.

	NTRU	RSA	McEliece
Encryption Speed	N^2	$N^2 \approx N^3$	N^2
Decryption Speed	N^2	N^3	N^2
Public-Key	N	N	N^2
Secret-Key	N	N	N^2
Message Expansion	$\log_p q - 1$	$1 - 1$	$1 - 1.6$

Table 10.2. Figure Comparison among NTRU, RSA and McEliece

10.5 Quantum Cryptography

It is evident that if a practical quantum computer is available, then all public-key cryptographic systems based on the difficulty of IFP, DLP, and ECDLP will be insecure. However, the cryptographic systems based on quantum mechanics will still be secure even if a quantum computer is available. In this section some basic ideas of quantum cryptography are introduced. More specifically, a quantum analog of the Diffie-Hellman key exchange/distribution system, proposed by Bennett and Brassard in 1984 [25], will be addressed.

First let us define four *polarizations* as follows:

$$\{0^\circ,\ 45^\circ,\ 90^\circ,\ 135^\circ\} \stackrel{\text{def}}{=} \{\rightarrow,\ \nearrow,\ \uparrow,\ \nwarrow\}. \tag{10.3}$$

The quantum system consists of a transmitter, a receiver, and a quantum channel through which polarized photons can be sent [26]. By the law of quantum mechanics, the receiver can either distinguish between the *rectilinear polarizations* $\{\rightarrow,\ \uparrow\}$, or reconfigure to discriminate between the diagonal polarizations $\{\nearrow,\ \nwarrow\}$, but in any case, he cannot distinguish both types. The system works in the following way:

[1] Alice uses the transmitter to send Bob a sequence of photons, each of them should be in one of the four polarizations $\{\rightarrow,\ \nearrow,\ \uparrow,\ \nwarrow\}$. For instance, Alice could choose, at random, the following photons

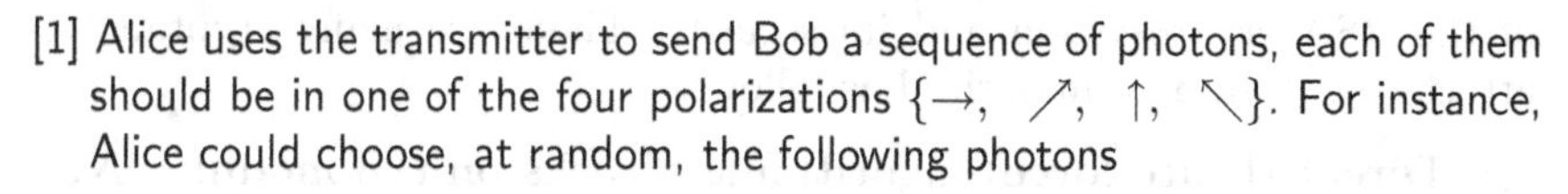

to be sent to Bob.

[2] Bob then uses the receiver to measure the polarizations. For each photon received from Alice, Bob chooses, at random, the following type of measurements $\{+,\ \times\}$:

$$+ \quad + \quad \times \quad \times \quad + \quad \times \quad \times \quad \times \quad +$$

[3] Bob records the result of his measurements but keeps it secret:

$$\uparrow \quad \rightarrow \quad \nearrow \quad \nwarrow \quad \rightarrow \quad \nearrow \quad \nearrow \quad \nearrow \quad \uparrow$$

[4] Bob publicly announces the type of measurements he made, and Alice tells him which measurements were of correct type:

$$\surd \qquad\qquad\qquad \surd \quad \surd \qquad\qquad \surd \qquad\qquad \surd$$

[5] Alice and Bob keep all cases in which Bob measured the correct type. These cases are then translated into bits $\{0, 1\}$ and thereby become the key:

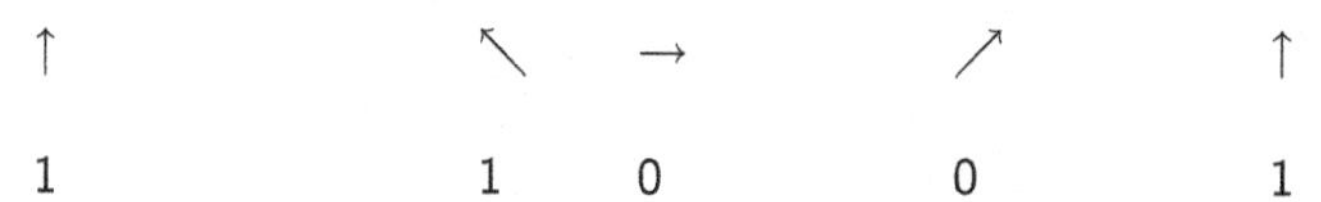

[6] Using this secret key formed by the quantum channel, Bob and Alice can now encrypt and send their ordinary messages via the classic public-key channel.

An eavesdropper is free to try to measure the photons in the quantum channel, but, according to the law of quantum mechanics, he cannot in general do this without disturbing them, and hence, the key formed by the quantum channel is secure.

10.6 Conclusions

On the display window of the MIT history and achievements on the wall of the ground floor of the MIT Building 10 in Cambridge, Massachusetts, there is a sentence to praise the RSA cryptosystem: *Ronald Rivest, Adi Shamir and Leonard Adleman invented the first workable public key cryptographic system, based on the use of very large prime numbers, that has so far proved unbreakable.* This remarkable brief sentence gives a very accurate description of the RSA cryptosystem and its security. There are many cryptanalytic attacks on RSA, as summarized as follows:

(1) **Direct (brute force) algorithmic attacks on the modulus N:**

- (1-1) Integer factorization attacks,
- (1-2) Discrete logarithm attacks,
- (1-3) Quantum factoring and discrete logarithm attacks,

  ......

(2) **Indirect (weakness) algorithmic attacks on $e, d, p, q, \phi(N), N$:**

- (2-1) Common modulus attack,
- (2-2) Fixed-point attacks,
- (2-3) Guessing d attacks,
- (2-4) Knowing $\phi(N)$ attack,
- (2-5) Forward attack,
- (2-6) e^{th} root attack,
- (2-7) Short e attacks,
- (2-8) Short d attacks,
- (2-9) Partial key exposure attacks,

  ......

(3) **Implementation (side-channel) attacks on d, p, q:**

(3-1) Timing attacks,

(3-2) Power attacks,

(3-3) Electromagnetic radiation attacks,

(3-4) Random fault (glitch) attacks,

$\cdots\cdots$

Remarkably enough, there are also many anti-attacks (defenses) against these attacks, that make RSA still useful, practical and conditionally unbreakable for 30 years since its invention in 1977. Nevertheless, RSA may become impractical if the RSA modulus N becomes too large. RSA may even become completely useless if an efficient integer factoring algorithm was developed. Thus, according to Bill Gates' advice (page 266 of [117]), we need to make an immediate transition to an alternative technique. Fortunately, there are many alternatives to replace RSA, the following are just some of these alternatives:

(1) If the size of the RSA modulus N is manageable, just use RSA, or, otherwise, use ECC, as for the same level of security, the key size of ECC will be significantly shorter than that of RSA.

(2) If a practical quantum computer becomes available, do not use either RSA or ECC anymore. Instead, use a quantum resistent cryptographic system, such as the lattice based cryptosystems ([144], [62], and [253]), or, the coding-based cryptosystem [203], discussed in this chapter.

(3) Use a quantum cryptographic system against the practical quantum computers. That is, in the quantum computer world, use quantum cryptography.

We end the book by a quotation of Sir Winston Churchill (1874-1965), a former British Prime Minister:

> *This is not the end. It is not even the beginning of the end. But it is, perhaps, the end of the beginning.*

Cryptography and cryptanalysis are such interesting areas that if old cryptosystems are broken, new ideas and new systems will always come up. No one can precisely predicate the future of cryptography, as Niels Bohr (the 1922 Nobel Prize Laureate of Physics) mentioned:

> *Prediction is very difficult, especially of the future.*

But it is no need to worry about the future. The feature should be always bright although the road towards the future cannot be smooth.

10.7 Chapter Notes and Further Reading

This chapter presented a very brief account of some of the possible replacements to the RSA cryptosystem just in case RSA becomes insecure duo to the progress in practical quantum computing, efficient factoring or some other unexpected attacks on RSA. For more information about elliptic curve cryptosystems, readers are suggested to consult Demytko [98], Hankerson et al [135], Koblitz [170] Menezes and Vanstone [206], Meyer and Müller [211], and Miller [213]. For code-based cryptography, see Berlekamp et al [29], Canteaut and Sendrier [63], Niederreiter [203], McEliece [228], and van Tilburg [318]. Quantum cryptography is currently a very hot research topic and readers may find the following reference helpful: Bennett [23], Bennett and Brassard [25], Bennett et al [26] Bruss et al [59], Lo [192], Lo and Chau [193], and Nielson and Chuang [229]. Finally, the series of conference proceedings entitled *Advances in Cryptology – Crypto, EuroCrypto, AsiaCrypto* published in Lecture Notes in Computer Science by Springer is an important source of new developments in all areas of cryptography and cryptanalysis.

Bibliography

1. O. Aciiçmez, "Yet Another Micro-Architectural Attack: Exploiting I-Cache", *Cryptology ePrint Archive*, Report 2007/164, May 2007.
2. O. Aciiçmez, S. Gueron and J. P. Seifert, "New Branch Prediction Vulnerabilities in OpenSSL and Necessary Software Countermeasures", *Cryptology ePrint Archive*, Report 2007/039, February 2007.
3. O. Aciiçmez, J. P. Seifert and Ç. K. Koç, "Micro-Architectural Cryptanalysis", *IEEE Security & Privacy*, July/August 2007, pp 62–64.
4. L. M. Adleman, "A Subexponential Algorithmic for the Discrete Logarithm Problem with Applications to Cryptography", *Proceedings of the 20th Annual IEEE Symposium on Foundations of Computer Science*, IEEE Press, 1979, pp 55–60.
5. L. M. Adleman, "Algorithmic Number Theory – The Complexity Contribution", *Proceedings of the 35th Annual IEEE Symposium on Foundations of Computer Science*, IEEE Press, 1994, pp 88–113.
6. L. M. Adleman, C. Pomerance, and R. S. Rumely, "On Distinguishing Prime Numbers from Composite Numbers", *Annals of Mathematics*, **117**, 1(1983), pp 173–206.
7. M. Agrawal, N. Kayal and N. Saxena, "Primes is in P", *Annals of Mathematics*, **160**, 2(2004), pp 781–793.
8. D. Agrawl, B. Archambeault, S. Chari and A. R. Rao, "Advances in Side-Channel Cryptanalysis: Electromagnetic Analysis and Template Attacks", *RSA Laboratories Cryptobytes*, **6**, (1)2003, pp 20–32.
9. M. Ajtai and C. Dwork, "A Pubic-Key Cryptosystem with Worst-Case/Average-Case Equivalence", *Proceedings of Annual ACM Symposium on Theory of Computing*, 1997, pp 284–293.
10. S. Alaca and H. S. Williams, *Introduction to Algebraic Number Theory*, Cambrdige University Press, 2004.
11. W. B. Alexi, B. Chor, O. Goldreich, and C. P. Schnorr, "RSA/Rabin Bits are $1/2+1/\text{poly}(\log N)$) Secure", *Proceedings of the 25th Annual IEEE Symposium on Foundations of Computer Science*, IEEE Press, 1984, pp 449–457.
12. W. B. Alexi, B. Chor, O. Goldreich, and C. P. Schnorr, "RSA/Rabin Function: Certian Parts are as Hard as the Whole", *SIAM Journal on Computing*, **17**, 2(1988), pp 194–209.
13. T. M. Apostol, *Introduction to Analytic Number Theory*, Corrected 5th Printing, Undergraduate Texts in Mathematics, Springer-Verlag, 1998.
14. A. O. L. Atkin and F. Morain, "Elliptic Curves and Primality Proving", *Mathematics of Computation*, **61**, (1993), pp 29–68.

15. D. Atkins, M. Graff, A. K. Lenstra, P. C. Leyland, "The Magic Words are Squeamish Ossifrage", *Advances in Cryptology ASIACRYPT'94*, Lecture Notes in Computer Science **917**, (1995), pp 261–277.
16. E. Bach, M. Giesbrecht and J. McInnes, *The Complexity of Number Theoretical Algorithms*, Technical Report 247/91, Department of Computer Science, University of Toronto, 1991.
17. E. Bach and J. Shallit, *Algorithmic Number Theory I – Efficient Algorithms*, MIT Press, 1996.
18. A. Baker, *A Concise Introduction to the Theory of Numbers*, Cambridge University Press, 1984.
19. D. Balenson, "Privacy Enhancement for Internet Electronic Mail: Part III: Algorithms, Modes, and Identifiers", *IETF RFC 1423*, 1993.
20. Bateman and Diamond, *Analytic Number Theory*, World Scientific, 2004.
21. F. L. Bauer, *Decrypted Secrets – Methods and Maxims of Cryptology*, 3rd Edition, Springer-Verlag, 2002.
22. M. Bellare and P. Rogaway, "Optimal asymmetric encryption", *Advances in Cryptology EUROCRYPT'94*, Lecture Notes in Computer Science **950**, Springer, 1995, pp 92–111.
23. C. H. Bennett, "Quantum Cryptography using any two Nonorthogonal Sates", *Physics Review Letters*, **68**, 1992, pp 3121–3124.
24. C. H. Bennett, "Quantum Information and Computation", *Physics Today*, October 1995, pp24–30.
smallskip
25. C. H. Bennett and G. Brassard, "Quantum Cryptography: Public Key Distribution and Coin Tossing", *Proccedings of the IEEE International Conference on Computersm Systems and Singal Processing*, IEEE Press, 1984, pp 175–179.
26. C. H. Bennett, G. Brassard and A. K. Ekert, "Quantum Cryptography", *Scientific American*, October 1992, pp 26–33.
27. C. H. Bennett, "Strengths and Weakness of Quantum Computing", *SIAM Journal on Computing*, **26**, (5)1997, pp 1510–1523.
28. C. H. Bennett and D. P. DiVincenzo, "Quantum Information and Computation", *Nature*, **404**, 2000, pp 247–255.
29. E. R. Berlekampe, R. J. McEliece and H. van Tilburg, "On the Inherent Intractability of Certain Coding Problems", *IEEE Transaction on Information Theory*, **IT-24**, 1978, pp 384–386.
30. E. Bernstein and U. Vazirani, "Quantum Complexity Theory", *SIAM Journal on Computing*, **26**, 5(1997), pp 1411–1473.
31. D. J. Bernstein, *Proving Primality After Agrawal-Kayal-Saxena*, Dept of Mathematics, Statistics and Computer Science, The University of Illinois at Chicago, 25 Jan 2003.
32. D. Bleichenbacher, "Chosen Ciphertext Attacks Against Protocols based on the RSA Encryption Standard PKCS#1", *Advances in Cryptology – Crypto '98*, Lecture Notes in Computer Science **1462**, Springer, 1998, pp 1–12.
33. D. Bleichenbacher, W. Bosma and A. K. Lenstra, "Some Remarks on Lucas-Based Cryptosystems", *Advances in Cryptology – Crypto '95*, Lecture Notes in Computer Science **963**, Springer, 1995, pp 386–396.
34. J. Blömer and A May, "Low Secret Exponent RSA Revisited", *Cryptography and Lattices*, Lecture Notes in Computer Science **2146**, Springer, 2001, pp 4–19.

35. M. Blum and S. Goldwasser, "An Efficient Probabilistic Public-key Encryption Scheme that Hides all Partial Information", *Advances in Cryptology - CRYPTO '84*, Lecture Notes in Computer Science **196**, Springer, 1985, pp 289–302.

36. E. Bombieri, *Problems of the Millennium: The Riemann Hypothesis*, Institute for Advanced Study, Princeton, and Clay Mathematics Institute, Boston, 2001.

37. D. Boneh, "Twenty Years of Attacks on the RSA Cryptosystem", *Notices of the AMS*, **46** 2(1999), pp 203–213.

38. D. Boneh and G. Durfee, "Cryptanalysis of RSA with Private Key d less than $N^{0.292}$", *IEEE Transactions on Information Theory*, **46**, (2000), pp 1339–1349.

39. D. Boneh, G. Durfee and Y. Frankel, "An Attack of RSA Given a Fraction of the Private Key Bits", *Advances in Cryptology - Asiacrypt '98*, Lecture Notes in Computer Science **1514**, Springer, 1998, pp 25–34.

40. D. Boneh and R. Venkatesan, "Breaking RSA may not be Equivalent to Factoring", *Advances in Cryptology - Eurocrypt '98*, Lecture Notes in Computer Science **1233**, Springer, 1998, pp 59–71.

41. D. Boneh, A. Joux and P. Q. Nguyen, "Why Textbook ElGamal and RSA Encryption are Insecure", *Advances in Cryptology – Asiacrypt 2000*, Lecture Notes in Computer Science **1976**, Springer, 2000, pp 30–43.

42. A. Bortz, D. Boneh and P. Nangy, "Exposing Private Information by Timing Web Applications", *Proceedings of the 16th International World Wide Web Conference*, Banff, Alberta, 8-12 May 2007.

43. A. Boscher, R. Naciri and E. Prouff, "CRT RSA Algorithm Protected Against Fault Attacks", *Information Security Theory and Practices: Smart Cards, Mobile and Ubiquitous Computing Systems*, Lecture Notes in Computer Science **4462**, 2007, pp 229-243.

44. R. P. Brent, "An Improved Monte Carlo Factorization Algorithm", *BIT*, **20**, (1980), pp 176–184.

45. R. P. Brent, "Some Integer Factorization Algorithms using Elliptic Curves", *Australian Computer Science Communications*, **8** (1986), pp 149–163.

46. R. P. Brent, "Primality Testing and Integer Factorization", *Proceedings of Australian Academy of Science Annual General Meeting Symposium on the Role of Mathematics in Science*, Canberra, 1991, pp 14–26.

47. R. P. Brent, "Uses of Randomness in Computation", Report TR-CS-94-06, Computer Sciences Laboratory, Australian National University, 1994.

48. R. P. Brent, "Some parallel algorithms for integer factorisation", *Proc. Fifth International Euro-Par Conference (Toulouse, France, 1-3 Sept 1999)*, Lecture Notes in Computer Science **1685**, Springer, 1999, pp 1–22.

49. R. P. Brent, "Twenty years' analysis of the binary Euclidean algorithm", *Millenial Perspectives in Computer Science: Proceedings of the 1999 Oxford-Microsoft Symposium in honour of Professor Sir Antony Hoare*, Edited by J. Davies, A. W. Roscoe and J. Woodcock, Palgrave, New York, 2000, pp 41–53.

50. R. P. Brent, "Recent progress and prospects for integer factorisation algorithms", *Proc. COCOON 2000 (Sydney, July 2000)*, Lecture Notes in Computer Science **1858**, Springer-Verlag, 2000, pp 3–22.

51. R. P. Brent, "Uncertainty Can Be Better than Certainty: Some Algorithms for Primality Testing", *Proc. COCOON 2000 (Sydney, July 2000)*, Strachey Lecture, Oxford Computing Laboratory, 2 November 2004.

52. R. P. Brent, G. L. Cohen and H. J. J. te Riele, "Improved Techniques for Lower Bounds for Odd Perfect Numbers", *Mathematics of Computation*, **57**, 196(1991), pp 857–868.
53. D. M. Bressoud, *Factorization and Primality Testing*, Undergraduate Texts in Mathematics, Springer-Verlag, 1989.
54. D. Bressoud and S. Wagon, *A Course in Computational Number Theory*, Key College Publishing and Springer-Verlag, 2000.
55. E. F. Brickell, D. M. Gordon and K. S. McCurley, "Fast Exponentiation with Precomputation (Extended Abstract)", *Advances in Cryptology - EUROCRYPT '92*, Lecture Notes in Computer Science **658**, Springer-Verlag, 1992, pp 200–207.
56. E. Brier, B. Chevallier-Mames, M. Ciet and C. Clavier, "Why One Should Also Secure RSA Public Key Elements", *Cryptographic Hardware and Embedded Systems - CHES 2006*, Lecture Notes in Computer Science *4249*, Springer, 2006, pp 324–338.
57. D. R. L. Brown, *Breaking RSA May be as Difficult as Factoring*, Certicom Research, 2006.
58. D. Brumley and D. Boneh, "Remote Timing Attacks are Practical", *Proceedings of 12th Usenix Security Symposium*, Wahsington, D. C., 4-8 August 2003, pp 1–14.
59. D. Bruss, G. Erdélyi, T. Meyer, T. Riege and J. Rothe, "Quantum Cryptography: A Survey", *ACM Computing Surveys*, **39**, 2(2007), Article 6, pp 1–27.
60. , J. A. Buchmann, *Introduction to Cryptography*, Springer, 2001.
61. S. A. Burr (editor), *The Unreasonable Effectiveness of Number Theory*, Proceedings of Symposia in Applied Mathematics **46**, American Mathematical Society, 1992.
62. J. Y. Cai and T. W. Cusick, "A Lattice-Based Public-Key Cryptosystem", *Information and Computation*, **151**, 1-2(1999), pp 17–31.
63. E. F. Canteaut and N. Sendrier, "Cryptanalysis of the Original McEliece Cryptosystem", *Advances in Cryptology – AsiaCrypto'98*, Lecture Notes in Computer Science **1514**, Springer, 1989, pp 187–199.
64. J.W.S. Cassels, *An Introduction to the Geometry of Numbers*, Springer, 1997.
65. J. R. Chen, "On the representation of a larger even integer as the sum of a prime and the product of at most two primes", *Scientia Sinica* **16**, (1973), pp 157–176.
66. L. Childs, *A Concrete Introduction to Higher Algebra*, 2nd Edition, Springer-Verlag, 2000.
67. S. Chowla, "There Exists an Infinity of 3-Combinations of Primes in A. P.", *Proceedings of Lahore Philosophy Society*, **6**, 2(1944), pp 15–16.
68. I. L Change, R. Laflamme, P, Shor, and W. H. Zurek, "Quantum Computers, factoring, and Decoherence, *Science*, **270**, 1995, pp 1633–1635.
69. A. Church, "An Unsolved Problem of Elementary Number Theory" *The American Journal of Mathematics*, **58**, 1936, pp 345–363.
70. A. Church, "Book Review: On Computable Numbers, with an Application to the Entscheidungsproblem by Turing", *Journal of Symbolic Logic*, **2**, 1937, pp 42–43.
71. H. Cohen, *A Course in Computational Algebraic Number Theory*, Graduate Texts in Mathematics **138**, Springer-Verlag, 1993.

72. S. Cook, *The Complexity of Theorem-Proving Procedures, Proceedings of the 3rd Annual ACM Symposium on the Theory of Computing*, New York, 1971, pp 151–158.
73. S. Cook, *The P versus NP Problem*, University of Toronto, April, 2000. (Manuscript prepared for the Clay Mathematics Institute for the Millennium Prize Problems; revised in November 2000.)
74. S. Cook and N. Thapen, "The Strength of Replacement in Weak Arithmetic", *Nineteenth Annual IEEE Symposium on Logic in Computer Science (LICS 2004)*, IEEE Press, 2004, pp 256–264.
75. J. W. Cooley and J. W. Tukey, "An Algorithm for the Machine Calculation of Complex Fourier Series", *Mathematics of Computation*, **19** (1965), pp 297–301.
76. D. Coppersmith, "Finding a Small Root of a Univariate Modular Equation", *Advances in Cryptology – Eurocrypt'96*, Lecture Notes in Computer Science **1070**, Springer, 1996, 155–165.
77. D. Coppersmith, "Finding a Small Root of a Bivariate Integer Equation; Factoring with High Bits Known", *Advances in Cryptology – Eurocrypt'96*, Lecture Notes in Computer Science **1070**, Springer, 1996, 178–180.
78. D. Coppersmith, "Small Solutions to Polynomial Equations, and Low Exponent RSA Vulnerability", *Journal of Cryptology*, **10**, (1997), pp 233–260.
79. D. Coppersmith, "Finding a Small Root of a Bivariate Integer Equation; Factoring with High Bits Known", *Cryptography and Lattices – CaLC 2001*, Lecture Notes in Computer Science **2146**, Springer, 2001, 20–31.
80. D. Coppersmith, "Solving Low Degree Polynomials", *Presentation at Asiacrypt*, Taipei, Taiwan, 1 December 2003.
81. D. Coppersmith, M. Franklin, J. Patarin and R. Reiter, "Low-Exponent RSA with Related Messages", *Advances in Cryptology - Eurocrypt'96*, Lecture Notes in Computer Science **1070**, Springer, 1996, 1–9.
82. T. H. Cormen, C. E. Ceiserson and R. L. Rivest, *Introduction to Algorithms*, MIT Press, 1990.
83. J. S. Coron, "30 Years of Attacks agianst RSA". Université du Luxenbourg, 2007, `http://jscoron.fr/blog/wp-content/uploads/2007/05/rsasurvey.pdf`.
84. J. S. Coron, M. Joye, D. Naccache and P Paillier, "New Attacks on PKCS#1 v1.5 Encryption", *Advances in Cryptology - EUROCRYPT 2000*, Lecture Notes in Computer Science **1807**, Springer, 2000, pp 369–381.
85. J. S. Coron and A. May, "Deterministic Polynomial-Time Equivalence of Computing the RSA Secret Key and Factoring", *Journal of Cryptology*, **20**, (1) 2007, pp 39–50.
86. J. S. Coron, D. Naccache, Y. Desmedt, A. Odlyzko and J. P. Stern, "Index Calculation Attacks on RSA Signature and Encryption" *Designs, Codes and Cryptography*, **38**, (1) 2006, pp 41–53.
87. J. S. Coron, D. Naccache, J. P. Stern "On the Security of RSA Padding", *Advances in Cryptology - CRYPTO'99*, Lecture Notes in Computer Science *1666*, Springer, 1999, pp 1–18.
88. S. C. Countinho, "The Mathematics of Ciphers: Number Theory and RSA Cryptography", A K Peters, 1999.
89. D. A. Cox, *Primes of the Form* $x^2 + ny^2$, Wiley, 1989.

90. R. Crandall and C. Pomerance, *Prime Numbers – A Computational Perspective*, 2nd Edition, Springer-Verlag, 2005.

91. P. A. Crouch, and J. H. Dvenport, "Lattice Attacks on RSA-Encrypted IP and TCP", *Cryptography and Coding*, Lecture Notes in Computer Science **2260**, Springer, 2001, pp 329–338.

92. S. Dasgupta, C. Papadimitriou, and U. Vazirani, *Algorithms*, McGraw-Hill, 2007.

93. H. Davenport, *The Higher Arithmetic*, 7th Edition, Cambridge University Press, 1999.

94. J. A. Davis and D. B. Holdridge, "Factoring using the Quadratic Sieve Algorithm", *Advances in Crypotology Crypto '83*, Plenum Press, 1984, pp 103–113.

95. J. F. Dhem, F. K. Koeune, P. A. Leroux, P. Mestré, J.J. Quisquater, and J. L. Willems, "Apractical Implementation of the Timing Attacks", *Smart Card: Research and Applications*, Lecture Notes in Computer Science **1820**, Springer, 2000, pp 167–182.

96. J. M. DeLaurentis, A Further Weakness in the Common Modulus Protocol for the RSA Cryptoalgorithm, *Crypotologia*, **8**, 1984, pp 253–259.

97. H. Delfs and H Knebl, *Introduction to Cryptography*, Springer, 2002.

98. N. Demytko, "A New Elliptic Curve Based Analogue of RSA", *Advances in Cryptology - EUROCRYPT '93*, Lecture Notes in Computer Science **765**, Springer, 1994, pp 40–49.

99. D. Deutsch, "Quantum Theory, the Church–Turing Principle and the Universal Quantum Computer", Proceedings of the Royal Society of London, Series **A**, **400**, (1985), pp 96–117.

100. L. E. Dickson, *History of the Theory of Numbers I – Divisibility and Primality*, G. E. Stechert & Co., New York, 1934.

101. W. Diffie and E. Hellman, "New Directions in Cryptography", *IEEE Transactions on Information Theory*, **22**, 5(1976), pp 644–654.

102. W. Diffie and E. Hellman, "Privacy and Authentication: An Introduction to Cryptography", *Proceedings of the IEEE*, **67**, 3(1979), pp 393–427.

103. P. G. L. Dirichlet, *Lecturers on Number Theory*, Supplements by R. Dedekind, American Mathematics Society and London Mathematics Society, 1999.

104. J. D. Dixon, "Factorization and Primality tests", *The American Mathematical Monthly*, June-July 1984, pp 333–352.

105. E. H. Edwards, *Riemann's Zeta Function*, Dover, 1974.

106. T. ElGamal, "A Public Key Cryptosystem and a Signature Scheme based on Discrete Logarithms", *IEEE Transactions on Information Theory*, **31** (1985), pp 496–472.

107. M. Ernst, E. Jochemsz, A. May and B. de Weger, "Partial Key Exposure Attacks on RSA up to Full Size Exponents", *Advances in Cryptology EUROCRYPT 2005*, Lecture Notes in Computer Science **3494**, Springer, 2005, pp 371–386.

108. Euclid, *The Thirteen Books of Euclid's Elements*, Translated by T. L. Heath, *Great Books of the Western World* **11**, edited by R. M. Hutchins, William Benton Publishers, 1952.

109. Euclid, *The Thirteen Books of Euclid's Elements*, Second Edition, Translated by Thomas L. Heath, Dover Publications, 1956.

110. Everest and Ward, *An Introduction to Number Theory*, Springer, 2005.

111. R. P. Feynman, "Simulating Physics with Computers", *International Journal of Theoretical Physics*, **21**, (1982), 467–488.

112. R. P. Feynman, *Feynman Lectures on Computation*, Edited by A. J. G. Hey and R. W. Allen, Addison-Wesley, 1996.

113. P. Fouque, S. Kunz-Jacques, G. Martinet, F. Muller and F. Valette, "Power Attack on Small RSA Public Exponent", *Cryptographic Hardare and Embedded Systems – CHES 2006*, Lecture Notes in Computer Science **4249**, Springer, 2006, pp 339–353.

114. E. Fujisaki, T. Okamoto, D. Pointcheval and J. Stern, "RSA – OAEP is secure under the RSA assumption", *Advances in Cryptology – CRYPTO 2001*, Lecture Notes in Computer Science **2139**, Springer, 2001, pp 260–274.

115. M. Gardner, "Mathematical Games – A New Kind of Cipher that Would Take Millions of Years to Break", *Scientific American*, **237**, 2(1977), pp 120–124.

116. M. R. Garey and D. S. Johnson, *Computers and Intractability – A Guide to the Theory of NP-Completeness*, W. H. Freeman and Company, 1979.

117. B. Gates, *The Road Ahead*, Viking, 1995.

118. C. F. Gauss, *Disquisitiones Arithmeticae*, G. Fleischer, Leipzig, 1801. English translation by A. A. Clarke, Yale University Press, 1966. Revised English translation by W. C. Waterhouse, Springer-Verlag, 1975.

119. N. Gershenfield and I. L. Chuang, "Bulk Spin-Resonance Quantum Computation", *Science*, **275**, 1997, pp 350–356.

120. M. Girault, R. Cohen and M. Campana, "A Generalized Birthday Attack", *Advances in Cryptology - EUROCRYPT '88*, Lecture Notes in Computer Science **330**, Springer, 1988, pp 129–156.

121. S. Goldwasser, "The Search for Provably Secure CryptoSystems", *Cryptology and Computational Number Theory*, Proceedings of Symposia in Applied Mathematics, **42**, American Mathematics Society, 1989, pp 89–114.

122. S. Goldwasser and J. Kilian, "Almost All Primes Can be Quickly Certified", *Proceedings of the 18th ACM Symposium on Theory of Computing*, Berkeley, 1986, pp 316–329.

123. S. Goldwasser and J. Kilian, "Primality Testing Using Elliptic Curves", *Journal of ACM*, **46**, 4(1999), pp 450–472.

124. S. Goldwasser and S. Micali, "Probabilistic Encryption", *Journal of Computer and System Sciences*, **28**, (1984), pp 270–299.

125. S. Goldwasser, S. Micali and C. Rackoff, "The Knowledge Complexity of Interactive Proof System", *SIAM Journal on Computing*, **18**, (1989), pp 186–208.

126. D. M. Gordon and K. S. McCurley, "Massively Parallel Computation of Discrete Logarithms", *Advances in Cryptology - Crypto '92*, Lecture Notes in Computer Science **740**, Springer-Verlag, 1992, pp 312–323.

127. D. M. Gordon, "Discrete Logarithms in GF(p) using the Number Field Sieve", *SIAM Journal on Discrete Mathematics*, **6**, 1(1993), pp 124–138.

128. B. Green and T. Tao, "The Primes Contain Arbitrarily Long Arithmetic Progressions", *arXiv:math/0404188v5*, 9 Feb 2007, 56 pages. Submitted to *Annual of Mathematics* on 8 April 2004 and last revised 9 Feb 2007..

129. J. Grobchadl, "The Chinese Remainder Theorem and its Application in a High-speed RSA Crypto Chip", *Proceedings of the 16th Annual Computer Security Applications Conference (ACSAC'00)*, IEEE Press, 2000, pp 384–393.

130. Grustka,J. *Quantum Computing*, McGraw-Hill, 1999.

131. S. Gueron and J. P. Seifert, "Is It Wise to Publish Your Public RSA Keys?", *Fault Diagnosis and Tolerance in Cryptography*, Lecture Notes in Computer Science **4236**, Springer, 2006, pp 1–12.

132. F. Guterl, "Suddenly, Number Theory Makes Sense to Industry", *International Business Week*, 20 June 1994, pp 62–64.

133. M. Gysin and J. Seberry, "Generalised Cycling Attacks on RSA and Strong RSA Primes", *Information Security and Privacy*, Lecture Notes in Computer Science **1587**, Springer, 1999, pp 149-163.

134. G. Hachez and J. J. Quisquater, "Montgomery Exponentiation with no Final Subtractions: Improved Results", *Cryptographic Hardware and Embedded Systems - CHES 2000*, Lecture Notes in Computer Science **1965**, Springer, 2000, pp 91–100.

135. D. Hankerson, A. J. Menezes and S. Vanstone, *Guide to Elliptic Curve Cryptography*, Springer, 2004.

136. J. Hastäd, "Solving Simutaneous Modlar Equations of Low Degres", *SIAM Journal on Computing*, **17**, (1988), pp 336–341.

137. G. H. Hardy, *A Mathematician's Apology*, Cambridge University Press, 1940 (1979 Reprinting).

138. G. H. Hardy and E. M. Wright, *An Introduction to Theory of Numbers*, 5th Edition, Oxford University Press, 1979.

139. J. Hastäd, "On Using RSA with Low Exponent in a Public Key Network", *Advances in Cryptology - Crypto '85*, Lecturer Notes in Computer Science **218**, Springer, 1986, pp 403–408.

140. M. Hellman, "Private Communications", 2001–2003.

141. M. J. Hinke, "Small Private Exponent Partial Key-Exposure Attacks on Multiprime RSA" *Technical Report CACR 2005-16*, University of Waterloo, 2005.

142. M. J. Hinke, "Another Look at Small RSA Exponents" *CT-RSA*, Edited bt D. Pointcheval 2006, Lecture Notes in Computer Science **3860**, Springer, 2006, pp. 82–98.

143. J. Hoffstein, J. Pipher and J. H. Silverman, "A Ring-Based Public-Key Cryptosystem", *Algorithmic Number Theory ANTS-III*, Lecture Notes in Computer Science **1423**, Springer, 1998, pp 267–288.

144. J. Hoffstein, N. Howgrave-Graham, J. Pipher, J. H. Silverman and W. Whyte, "NTRUEncrypt and NTRUSign: Efficient Public Key Algorithmd for a Post-Quantum World", *Proceedings of the International Workshop on Post-Quantum Cryptography (PQCrypto 2006)*, 23-26 May 2006, pp 71–77.

145. J. Hopcroft, R. Motwani and J. Ullman, *Introduction to Automata Theory, Languages, and Computation*, 3rd Edition, Addison-Wesley, 2007.

146. N. Howgrave-Graham, "Finding Small Roots of Univariate Modular Equations Revisited", *ACrytography and Coding*, Lecture Notes in Computer Science **1355**, Springer, 1997, 131–142.

147. L. Hua, *Introduction to Number Theory*, English Translation from Chinese by P. Shiu, Springer-Verlag, 1980.

148. R. J. Hughes, "Cryptography, Quantum Computation and Trapped Ions", *Philosophic Transactions of the Royal Society London*, Series **A**, **356** (1998), pp 1853–1868.

149. R. M. Huizing, *An Implementation of the Number Field Sieve*, Note NM-R9511, Centre for Mathematics and Computer Science (CWI), Amsterdam, 1995.

150. D. Husemöller, *Elliptic Curves*, Graduate Texts in Mathematics **111**, Springer-Verlag, 1987.

151. H. Inamori, *A Minimal Introduction to Quantum Key Distribution*, Centre for Quantum Computation, Clarendon Laboratory, Oxford University, 1999.

152. A. E. Ingham, *The Distribution of Prime Numbers*, Forworded by R. C. Vaughan in 1990, Cambridge University Press, 1995.

153. K. Ireland and M. Rosen, *A Classical Introduction to Modern Number Theory*, 2nd Edition, Graduate Texts in Mathematics **84**, Springer-Verlag, 1990.

154. ITU-X Recommendation X.509, "Information technology - Open Systems Interconnection - The Directory: Authentication framework", 1989.

155. O. Jacobi and Y. Jacobi, "A New Related Message Attack on RSA", *Public Key Cryptography - PKC 2005*, Lecture Notes in Computer Science **3386**, Springer, 2005, pp 1–8.

156. M. J. Jacobson, N. Koblitz, J. H. Silverman, A. Stein, E. Teske, "Analysis of the Xedni Calculus Attack", *Designs, Codes and Cryptography*, **20**, 2000, pp 41-64.

157. E. Jochemsz and A. May, "A Polynomial Time Attack on RSA with Private CRT-Exponents Smaller Than $N^{0..73}$", *Advances in cryptology – Crypto 2007*, Lecture Notes in Computer Science **4622**, Springer, 2007, pp 395–411.

158. M. Joye and J. J. Quisquater, "Cryptanalysis of RSA-Type Cryptosystems: A Visit", *Networks Threats*, DIMACS Series in Discrete Mathematics and Theoretical Computer Science, American Mathematical Society, **38**, 1998, pp 21–31.

159. M. Joye, J. J. Quisquater and T.Takagi, "How to Choose Secret Parameters for RSA and its Extensions to Elliptic Curves", *Designs, Codes and Cryptography*, **23**, 2001, pp 297–316.

160. M. F. Jones, M. Lal and W. J. Blundon, "Statistics on Certain Large Primes", *Mathematics of Computation*, **21**, 97(1967), pp 103-107.

161. D. Kahn, *The Codebreakers: The Story of Secret Writing*, Macmillan, 1976

162. B. Kaliski and M. Robshaw, "The Secure Use of RSA", *CryptoByte*, **1**, (1995), 7-13.

163. B. Kaliski, "Timing Attacks on Cryptosystems", *RSA Laboratories Bulletin*, Number 2, January 1996.

164. R. Karp, "Reducibility among Cominatorial Problems", *Complexity of Computer Computations*, Edited by R. E. Miller and J. W. Thatcher, Plenum Press, New York, 1972, 85–103.

165. S. Katzenbeisser "Recent Advances in RSA Cryptography", Kluwer Academic Publishers, 2001.

166. J. Kilian, *Uses of Randomness in Algorithms and Protocols*, MIT Press, 1990.

167. C. H. Kim and J. J. Quisquater, "Fault Attacks for CRT Based RSA: New Attacks, New Results, and New Countermeasures", *Information Security Theory and Practices. Smart Cards, Mobile and Ubiquitous Computing Systems*, Lecture Notes in Computer Science **4462**, 2007, pp215–228.

168. V. Klma and T Rosa, "Further Results and Considerations on Side Channel Attacks on RSA" Lecture Notes in Computer Science **2523**, Springer, 2003, pp 63–92.

169. D. E. Knuth, *The Art of Computer Programming II – Seminumerical Algorithms*, 3rd Edition, Addison-Wesley, 1998.

170. N. Koblitz, "Elliptic Curve Cryptography", *Mathematics of Computation*, **48** (1987), 203–209.

171. N. Koblitz, *Introduction to Elliptic Curves and Modular Forms*, 2nd Edition, Graduate Texts in Mathematics **97**, Springer-Verlag, 1993.

172. N. Koblitz, *A Course in Number Theory and Cryptography*, 2nd Edition, Graduate Texts in Mathematics **114**, Springer-Verlag, 1994.

173. N. Koblitz, *Algebraic Aspects of Cryptography*, Algorithms and Computation in Mathematics **3**, Springer-Verlag, 1998.

174. N. Koblitz, "A Survey of Number Theory and Cryptography", *Number Theory*, Edited by . P. Bambah, V. C. Dumir and R. J. Hans-Gill, Birkhäser, 2000, pp 217–239.

175. N. Koblitz, "Cryptography", *Mathematics Unlimited – 2001 and Beyond*, Edited by B. Enguist and W. Schmid, Springer-Verlag, 2001, pp 749–769.

176. N. Koblitz and A. J. Menezes, "A Survey of Public-Key Cryptosystems", *SIAM Review*, **46**, 2004, pp 599–634.

177. N. Koblitz and A. J. Menezes, "Another look at 'provable security',", *Journal of Cryptology*, **20**, 2004, pp 3–37.

178. P. Kocher, "Timing attacks on implementations of Diffie-Hellman, RSA, DSS, and other systems", *Advances in Cryptology crypto '96*, Springer-Verlag, LNCS 1109, 1996, pp 104–113.

179. A. G. Konheim, *Computer Security and Cryptography*, Wiley, 2007.

180. K. Koyama, U. U. Maurer, T. Okamoto and S. A. Vanstone, "New Public-Key Scheme based on Elliptic Curve over the Ring $\mathbb{Z}_n$", *Advances in Cryptology - CRYPTO '91*, Lecture Notes in Computer Science **576**, Springer, 1992, pp 252-266.

181. U. Kühn, "Side-Channel Attacks on Textbook RSA and ELGamal Encryption", *Proceedings of the Public-Key Cryptography (PKC 2003)*, Lecture Notes in Computer Science **2567**, Springer, 2003, pp 324–336.

182. K. Kurosawa, K. Okada and Shigeo Tsujii, "Low exponent attack against elliptic curve RSA", *Advances in Cryptology ASIACRYPT '94*, Lecture Notes in Computer Science **917**, Springer, 1995, pp 376–383.

183. L. J. Lander and T. R. Parkin, "Consecutive primes in arithmetic progression", *Mathematics of Computation*, **21**, 99(1967), page 489.

184. D. H. Lehman, "A Theorem in the Theory of Numbers", *Bulletin of the American Mathematical Society*, **14**, (1907), pp 501–502.

185. R. S. Lehman, "Factoring Large Integers", *Mathematics of Computation*, **28**, 126(1974), pp 637–646.

186. H. W. Lenstra, Jr., "Factoring Integers with Elliptic Curves", *Annals of Mathematics*, **126**, (1987), pp 649–673.

187. A. K. Lenstra and H. W. Lenstra, Jr. (editors), *The Development of the Number Field Sieve*, Lecture Notes in Mathematics **1554**, Springer-Verlag, 1993.

188. A. K. Lenstra, H. W. Lenstra, Jr., and L. Lovász, "Factoring Polynomials with Rational Coefficients", *Mathematische Annalen*, **261**, (1982), pp 515–534.

189. A. K. Lenstra and R. R. Verheul, "Selecting Cryptographic Key Sizes", *Journal of Cryptology*, **14**, 2001, pp 255–293.

190. W. J. LeVeque, *Fundamentals of Number Theory*, Dover, 1977.

191. H. R. Lewis and C. H. Papadimitrou, *Elements of the Theory of Computation*, Prentice-Hall, 1998.

192. H. K. Lo, "Quantum Cryptography", *Introduction to Quantum Computation and Information*, edited by H. K. Lo, S. Popescu and T. Spiller, World Scientific, 1998, 76–119.

193. H. Lo and H. Chau, "Unconditional Security of Quantum key Distribution over Arbitrary Long Distances", *Science*, **283**, 1999, 2050–2056.

194. F. J. MacWilliams and N. J. A. Sloana, *Power Analysis Attacks: Revealing the Secrets of Smart Cards*, Springer, 2007.

195. S. Mangard, E. Oswald and T. Popp, *Power Analysis Attacks: Revealing the Secrets of Smart Cards*, Springer, 2007.

196. W. Mao, *Modern Cryptography*, Prentice-Hall, 2004.

197. K. Martin, "Secure Communication without Encryption', *IEEE Security & Privacy*, **5**, 2(2007), pp 68–71.

198. R. Mayer-Sommer, "Smartly Analyzing the Simplicity and the Power of Simple Power Analysis on Smartcards", *Cryptographic Hardware and Embedded Systems - CHES 2000*, Lecture Notes in Computer Science **1965**, Springer, 2000, pp 78–92.

199. S. McMath, *Daniel Shanks' Square Forms Factorization*, United States Naval Academy, Annapolis, Maryland, 24 November 2004.

200. D. C. Marinescu and G. M. Marinescu, *Approaching Quantum Computing*, Prentice-Hall, 2005.

201. K. S. McCurley, "The Discrete Logarithm Problem", *Cryptology and Computational Number Theory*, edited by C. Pomerance, Proceedings of Symposia in Applied Mathematics **42**, American Mathematics Society, 1990, pp 49–74.

202. K. S. McCurley, "Odds and Ends from Cryptology and Computational Number Theory", *Cryptology and Computational Number Theory*, edited by C. Pomerance, Proceedings of Symposia in Applied Mathematics **42**, American Mathematics Society, 1990, pp 49–74.

203. R. J. McEliece, *A Public-Key Cryptosystem based on Algebraic Coding Theory*, JPL DSN Progress Report 42-44, 1978, pp 583–584.

204. A. R. Meijer, "Groups, Factoring, and Cryptography" *Mathematics Magazine*, **69**, 2(1996), pp 103–109.

205. S. Müller and W. B. Müller, "The Security of Pubic-Key Cryptosystems based on Integer Factorization", *Information Security and Privacy - ACISP 1998*, Lecture Notes in Computer Science **1438**, Springer, 1998, pp 9–23.

206. A. Menezes and S. A. Vanstone, "Elliptic curve cryptosystems and their implementation", *Journal of Cryptology*, **6**, (1993), pp 209–224.

207. A. Menezes, P. C. van Oorschot and S. A. Vanstone, *Handbook of Applied Cryptosystems*, CRC Press, 1996.

208. R. C. Merkle, "Secure Communications over Insecure Channels" *Communications of the ACM*, **21**, (1978), pp 294–299. (Submitted in 1975.)

209. T. S. Messerges, E. A. Dabbish and R. H. Sloan, "Power Analysis Attack on Modular Exponentiation in Smartcard", *Cryptographic Hardare and Embedded Systems – CHES 1999*, Lecture Notes in Computer Science **1717**, Springer, 1999, pp 144–157.

210. J. F. Mestre, "Formules Explicites et Minoration de Conducteurs de Variétés algébriques" *Compositio Mathematica*, **58**, (1986), pp 209–232.

211. B. Meyer and V. Müller, "A Public Key Cryptosystem Based on Elliptic Curves over $\mathbb{Z}/n\mathbb{Z}$ Equivalent to Factoring", *Advances in Cryptology*, EUROCRYPT '96, Proceedings, Lecture Notes in Computer Science **1070**, Springer-Verlag, 1996, pp 49–59.

212. G. Miller, "Riemann's Hypothesis and Tests for Primality", *Journal of Systems and Computer Science*, **13**, (1976), pp 300–317.

213. V. Miller, "Uses of Elliptic Curves in Cryptography", *Advances in Cryptology*, CRYPTO '85, Proceedings, Lecture Notes in Computer Science **218**, Springer-Verlag, 1986, 417–426.

214. J. F. Misarsky, "A Multiplicative Attack using LLL Algorithm on RSA Signatures with Redundancy", *Advances in Cryptology CRYPTO '97*, Lecture Notes in Computer Science **1294**, 1997, pp 221–234.

215. R. A. Mollin, *Algebraic Number Theory*, Chapman & Hall/CRC, 1999.

216. R. A. Mollin, *An Introduction to Cryptography*, Chapman & Hall/CRC, 2001.

217. R. A. Mollin, *RSA and Public-Key Cryptography*, Chapman & Hall/CRC Press, 2003.

218. R. A. Mollin, *Codes: The Guide to Secrecy from Ancient to Modern Times*, Chapman & Hall/CRC Press, 2005.

219. P. L. Montgomery, "Multiplication without Trial Division", *Mathematics of Computation*, **44**, (1985), pp 519–521.

220. P. L. Montgomery, "Speeding the Pollard and Elliptic Curve Methods of Factorization", *Mathematics of Computation*, **48**, 177(1987), pp 243–264.

221. P. L. Montgomery, "A Survey of Modern Integer Factorization Algorithms", *CWI Quarterly*, **7**, 4(1994), pp 337–394.

222. F. Morain, "Implementing the Asymptotically Fast Version of the Elliptic Curve Primality Proving Algorithm", *Mathematics of Computation*, **76** (2007), pp 493–505.

223. M. A. Morrison and J. Brillhart, "A Method of Factoring and the Factorization of F_7", *Mathematics of Computation*, **29** (1975), pp 183–205.

224. R. Motwani and P. Raghavan, *Randomized Algorithms*, Cambridge University Press, 1995.

225. S. Miller, "On the Security of an RSA Based Encryption Scheme", *Information Security and Privacy: 4th Australasian Conference, ACISP'99*, Lecture Notes in Computer Science, **1587**, 1999, pp 135–148.

226. D. Naccache and M. Tunstall, "How to Explain Side-Channel Leakage to Your Kids", *Cryptographic Hardware and Embedded Systems - CHES 2000*, Lecture Notes in Computer Science **1965**, Springer, 2000, pp 49–80.

227. M. B. Nathanson, *Elementary Methods in Number Theory*, Springer-Verlag, 2000.

228. H. Niederreiter, "Knapsack Type Cryptosystems and Algebraic Coding Theory", *Problem of Control and Information Theory*, **15**, 1986, pp 159–166.

229. M. A. Nielson and I. L. Chuang, *Quantum Computation and Quantum Information*, Cambridge University Press, 2000.

230. D. Neuenschwander, "RSA and Probabilistic Prime Number Tests", *Probabilistic and Statistical Methods in Cryptology*, Lecture Notes in Computer Science **3028**, Springer, 2004, 17–35.

231. I. Niven, H. S. Zuckerman and H. L. Montgomery, *An Introduction to the Theory of Numbers*, 5th Edition, John Wiley & Sons, 1991.

232. A. M. Odlyzko, "Discrete Logarithms in Finite Fields and their Cryptographic Significance", *Advances in Cryptography*, EUROCRYPT '84, Proceedings, Lecture Notes in Computer Science **209**, Springer, 1984, pp 225–314.

233. A. M. Odlyzko, "Discrete Logarithms: the Past and the future", *Design, Codes, and Cryptography*, **19**, (2000), pp 129–145.

234. C. H. Papadimitrou, *Computational Complexity*, Addison Wesley, 1994.

235. S. J. Patterson, *An Introduction to the Theory of the Riemann Zeta-Function*, Cambridge University Press, 1988.

236. R. G. E. Pinch, "Some Primality Testing Algorithms", *Notices of the American Mathematical Society*, **40**, 9(1993), pp 1203–1210.

237. R. G. E. Pinch, "Extending the Wiener Attack to RSA-Type Cryptosystem", *Electronic Letters*, **31**, (20) 1995, pp 1736–1738.

238. R. G. E. Pinch, *Mathematics for Cryptography*, Queen's College, University of Cambridge, 1997.

239. A. O. Pittenger, *An Introduction to Quantum Comuting Algorithms*, Birkhäuser, 2001.

240. B. Pfitzmann and M. Waidner, "Attacks on Protocols for Server-Aided RSA Computation", *Advances in Cryptology - EUROCRYPT '92*, Lecture Notes in Computer Science **658**, Springer, 1993, pp 153–162.

241. S. C. Pohlig and M. Hellman, "An Improved Algorithm for Computing Logarithms over GF(p) and its Cryptographic Significance", *IEEE Transactions on Information Theory*, **24** (1978), pp 106–110.

242. J. M. Pollard, "Theorems on Factorization anf Primality Testing", *Procedings of Cambridge Philosophy Society*, **76** (1974), pp 521–528.

243. J. M. Pollard, "A Monte Carlo Method for Factorization", *BIT*, **15**, (1975), pp 331–332.

244. J. M. Pollard, "Monte Carlo Methods for Index Computation (mod p)", *Mathematics of Computation*, **32**, (1980), pp 918–924.

245. C. Pomerance, "Analysis and Comparison of some Integer Factoring Algorithms", in H. W. Lenstra, Jr. and R. Tijdeman, eds., *Computational Methods in Number Theory*, No **154/155**, Mathematical Centrum, Amsterdam, 1982, pp 89–139.

246. C. Pomerance, "The Quadratic Sieve Factoring Algorithm", *Proceedings of Eurocrypt 84*, Lecture Notes in Computer Science **209**, Springer, 1985, pp 169–182.

247. C. Pomerance (editor), "Cryptology and Computational Number Theory – An Introduction", *Proceedings of Symposia in Applied Mathematics* **42**, American Mathematical Society, 1990.

248. C. Pomerance, "A Tale of Two Sieves", *Notice of the AMS*, **43**, 12(1996), pp 1473–1485.

249. "Elementary Thoughts on Discrete Logarithms", Proceedings of a MSRI Workshop, edited by J. Buhler and P. Stevenhage, 2004.

250. J. Proos and C. Zalka, "Shor's Discrte Logarithm Quantum Algorithm for Elliptic Curves", Dept of Combinatorics and Optimization, University of Waterloo, 25 January 2003, 35 pages.

251. M. O. Rabin, "Probabilistic Algorithms for Testing Primality", *Journal of Number Theory*, **12**, (1980), pp 128–138.

252. M. Rabin, "Digitalized Signatures and Public-Key Functions as Intractable as Factorization", *Technical Report MIT/LCS/TR-212*, MIT Laboratory for Computer Science, 1979.

253. O. Regev, "New Lattice Based Cryptographic Constructions", *Proceedings of the 35th Annual ACM Symposium on the Theory of Computing*, San Diego, California, 2003, pp 407–416.

254. P. Ribenboim, *The Little Book on Big Primes*, Springer-Verlag, 1991.

255. P. Ribenboim, "Selling Primes", *Mathematics Magazine*, **68**, 3(1995), pp 175–182.

256. P. Ribenboim, *The New Book of Prime Number Records*, Springer-Verlag, 1996.

257. H. J. J. te Riele, "Factorization of RSA-140 using the Number Field Sieve", `http://www.crypto-world.com/announcements/RSA140.txt`, 4 February 1999.

258. H. J. J. te Riele, "Factorization of a 512-bits RSA Key using the Number Field Sieve", `http://www.crypto-world.com/announcements/RSA155.txt`, 26 August 1999.

259. H. J. J. te Riele, W. Lioen and D. Winter, "Factoring with the Quadrtaic Sieve on Large Vector Computers", *Journal of Computational and Applied Mathematics*, **27**, (1989), pp 267–278.

260. H. Riesel, *Prime Numbers and Computer Methods for Factorization*, Birkhäuser, Boston, 1990.

261. R. L. Rivest, A. Shamir and L. Adleman, *On Digital Signatures and Public Key Cryptosystems*, Technical Memo 82, Laboratory for Computer Science, Massachusetts Institute of Technology, April 1977.

262. R. L. Rivest, A. Shamir and L. Adleman, "A Method for Obtaining Digital Signatures and Public Key Cryptosystems", *Communications of the ACM*, **21**, 2(1978), pp 120–126.

263. R. L. Rivest, "Remarks on a Proposed Cryptanalytic Attack on the M.I.T. Public-key Cryptosystem", *Cryptologia*, **2**, 1(1978), pp 62–65.

264. R. L. Rivest and B. Kaliski, RSA Problem, In: *Encyclopedia of Cryptography and Security*, Edited by H. C. A. van Tilborg, Springer-Verlag, 2005.

265. R. L. Rivest and A. Shamir, "Efficient Factoring based on Partial Information", *Advances in Cryptology - EUROCRYPT '86*, Lecture Notes in Computer Science **219**, Springer, 1986, pp 31–34.

266. H. E. Rose, *A Course in Number Theory*, 2nd Edition, Oxford University Press, 1994.

267. K. Rosen, *Elementary Number Theory and its Applications*, 5th Edition, Addison-Wesley, 2005.

268. S. Ross, *A First Course in Probability*, 7th Edition, Prentice Hall, 2006.

269. J. Rothe, *Complexity Theory and Cryptography*, Springer, 2005.

270. J. J. Rotman *An Introduction to the Theory of Groups*, Springer-Verlag, 1994.

271. RSA Laboratories, "PKCS #1: RSA Encryption Standard", Version 1.4, 1993; Version 2.1, 2002.

272. Y. Sakai and K. Sakurai, "A New Attack with Side Channel Leakage During Exponent Recoding Computations", *Cryptographic Hardware and Embedded Systems - CHES 2004*, Lecture Notes in Computer Science **3156**, Springer, 2004, pp 298–311.

273. I. K. Salah, A. Darwish and A. Oqeili, "Mathematical Attacks on RSA Cryptosystem", *Journal of Computer Science*, **2** (8), 2006, pp665–671.

274. A. Salomaa, *Public-Key Cryptography*, 2nd Edition, Springer-Verlag, 1996.

275. W. Schindley, "A Combined Timing and Power Attack", *Proceedings of the Public-Key Cryptography (PKC 2002)*, Lecture Notes in Computer Science **2274**, Springer, 2002, pp 263–279.

276. O. Schirokauer, D. Weber, and T. Denny, "Discrete Logarithms, the Effectiveness of the Index Calulus Method", *Algorithmic Number Theory*, Lecture Notes in Computer Science **1122**, Springer, 1996, pp 337–361.

277. B. Schneier, *Applied Cryptography – Protocols, Algorithms, and Source Code in C*, 2nd Edition, John Wiley & Sons, 1996.

278. B. Schneier, "The Secret Story of Non-Secret Encryption", *Crypto-Gram Newsletter*, Counterpane Systems, May 15, 1998.

279. W. Schindler, "A Timing Attack against RSA with the Chinese Remainder Theorem ", *Cryptographic Hardware and Embedded Systems - CHES 2000*, Lecture Notes in Computer Science **1965**, Springer, 2000, pp 109–124.

280. W. Schindler and C. D. Walter, "More Detail for a Combined Timing and Power Attack against Implementations of RSA", *Cryptography and Coding*, Lecture Notes in Computer Science **2898**, Springer, 2003, 245–263.

281. C. P. Schnorr, "Efficient Identification and Signatures for Smart Cards", *Advances in Cryptography*, CRYPTO '89, Proceedings, Lecture Notes in Computer Science **435**, Springer, 1990, pp 239–252.

282. R. Schoof, "Elliptic Curves over Finite Fields and the Computation of Square Roots mod p", *Mathematics of Computation*, **44**, (1985), pp 483–494.

283. M. R. Schroeder, *Number Theory in Science and Communication*, 3rd Edition, Springer Series in Information Sciences **7**, Springer-Verlag, 1997.

284. A. Shamir, "How to Share a Secret", *Communications of the ACM*, **22**, 11(1979), pp 612–613.

285. C. Shannon, "Communication Theory of Secrecy Systems", *Bell System Technical Journal*, **28**, 1949, pp 656–715.

286. K. Shen, J. N. Crossley and A. W.-C. Lun, *The Nine Chapters on the Mathematical Art: Companion and Commentary*, Oxford University Press and Beijing Sicience Press, 1999.

287. P. Shor, "Algorithms for Quantum Computation: Discrete Logarithms and Factoring", *Proceedings of 35th Annual Symposium on Foundations of Computer Science*, IEEE Computer Society Press, 1994, pp 124–134.

288. P. Shor, "Polynomial-Time Algorithms for Prime Factorization and Discrete Logarithms on a Quantum Computer", *SIAM Journal on Computing*, **26**, 5(1997), pp 1484–1509.

289. T. O. Silva, "Goldbach Conjecture Verification", `http://www.ieeta.pt/~tos/goldbach.html`, 2007.

290. J. H. Silverman and J. Tate, *Rational Points on Elliptic Curves*, Undergraduate Texts in Mathematics, Springer-Verlag, 1992.

291. J. H. Silverman, *The Arithmetic of Elliptic Curves*, Graduate Texts in Mathematics **106**, Springer-Verlag, 1994.
292. J. H. Silverman, "The Xedni Calculus and the Elliptic Curve Discrete Logarithm Problem", *Designs, Codes and Cryptography*, **20**, 2000, pp 5-40.
293. J. H. Silverman, "The Xedni Calculus and the Elliptic Curve Discrete Logarithm Problem", *Designs, Codes and Cryptography*, **20**, 2000, pp 5-40.
294. J. H. Silverman, *A Friendly Introduction to Number Theory*, 3rd Edition, Prentice-Hall, 2006.
295. J. H. Silverman and J. Suzuki, "Elliptic Curve Discrete Logarithms and the Index Calculus", *Advances in Cryptology – ASIACRYPT '98*, Lecture Notes in Computer Science **1514**, Springer, 1998, pp 110–125.
296. R. D. Silverman, 'The Multiple Polynomial Quadratic Sieve", *Mathematics of Computation*, **48**, (1987), pp 329–339.
297. R. D. Silverman, "A Perspective on Computational Number Theory", *Notices of the American Mathematical Society*, **38**, 6(1991), pp 562–568.
298. R. D. Silverman, "Massively Distributed Computing and Factoring Large Integers", *Communications of the ACM*, **34**, 11(1991), pp 95–103.
299. D. R. Simon, "On the Power of Quantum Computation", *Proceedings of the 35th Annual IEEE Symposium on Foundations of Computer Science*, IEEE Press, 1994, pp 116–123.
300. G. J. Simons and M. J. and Morris, A Weakness Privacy Protocol using the RSA Crypto Algorithm, *Crypotologia*, **1**, 1977, pp 406–411.
301. G. J. Simons, A Weakness Privacy Protocol using the RSA Crypto Algorithm, *Crypotologia*, **1**, 1983, pp 180–182.
302. S. Singh, *The Code Book – The Science of Secrecy from Ancient Egypt to Quantum Cryptography*, Fourth Estate, London, 1999.
303. S. Singh, *The Science of Secrecy – The Histroy of Codes and Codebreaking*, Fourth Estate, London, 2000.
304. M. Sipser, *Introduction to the Theory of Computation*, 2nd Edition, Thomson, 2006.
305. P. Smith and M. Lennon, "LUC: A New Public-Key System", *Proceedings of Ninth International Conference on Information Security, IFIP/Sec '93*, Toronto, May 12-14, 1993, pp 103-117.
306. R. Steinfeld, S. Contini, H. Wang and J. Pieprzyk, "Converse Results to the Wiener Attack on RSA", *Public Key Cryptography - PKC 2005*, Lecture Notes in Computer Science **3386**, 2005, pp 184–198.
307. R. Solovay and V. Strassen, "A Fast Monte-Carlo Test for Primality", *SIAM Journal on Computing*, **6**, 1(1977), 84–85. "Erratum: A Fast Monte-Carlo Test for Primality", *SIAM Journal on Computing*, **7**, 1(1978), page 118.
308. J. Stillwell, *Elements of Number Theory*, Springer-Verlag, 2000.
309. V. Shoup, *A Computational Introduction to Number Theory and Algebra*, Cambridge University Press, 2005.
310. M. Stamp and R. M. Low, "Applied Cryptanalysis: Breaking Ciphers in the Real World", Wiley, 2007.
311. D. R. Stinson, *Cryptography: Theory and Practice*, 3rd Edition, Chapman & Hall/CRC Press, 2005.
312. Stopple, *A Primer of Analytic Number Theory – From Pythagoras to Riemann*, Cambridge University Press, 2003.

313. V. Strassen, "The Computational Complexity of Continued Fractions", *SIAM Journal on Computing*, **12**, (1) 1983, pp 1–27.

314. H. M. Sun and M. E. Wu, "An Approach Towards Rebalanced RSA-CRT with Short Public Exponent", *Cryptology ePrint Archive*, Report 2005/053, 2005. `http://eprint.iacr.org/`.

315. W. Trappe and L. Washington, *Introduction to Cryptography with Coding Theory*, 2nd Edition, Prentice-Hall, 2006.

316. A. Turing, "On Computable Numbers, with an Application to the Entscheidungsproblem", *Proceedings of the London Mathematical Society*, Series 2 **42**, pp 230–265 and **43**, pp 544–546.

317. L. M. K. Vandersypen, M. Steffen, G. Breyta, C. S. Tannoni, M. H. Sherwood, and I. L. Chuang, "Experiemental Realization of Shor's Quantum Factoring Algorithm Uisng Nuclear Magnetic Resonance", *Nature*, **414**, 20/27 December 2001, pp 883–887.

318. H. van Tilburg, "On the McEliece Public-Key Cryptography", *Advances in Cryptology – Crypto'88*, Lecture Notes in Computer Science **403**, Springer, 1989, pp 119–131.

319. H. van Tilborg, *Fundamentals of Cryptography*, Kluwer Academic Publishers, 1999.

320. H. van Tilborg (editor), *Encyclopedia of Cryptography and Security*, Springer, 2005.

321. S. S. Wagstaff, Jr., *Cryptanalysis of Number Theoretic Ciphers*, Chapman & Hall/CRC Press, 2002.

322. C. D. Walter, "Seeing through MIST Given a Small Fraction of an RSA Private Key", *Topics in Cryptology - CT-RSA 2003*, Lecture Notes in Computer Science **2612**, Springer, 2003, pp 391–402.

323. B. de Weger, "Cryptanalysis of RSA with Small Prime Difference", *Applicable Algebra in Engineering, Communication and Computing*, **13**, (1) 2002, pp 17–28.

324. H. Wiener, "Cryptanalysis of Short RSA Secret Exponents", *IEEE Transactions on Information Theory*, **36**, 3(1990), pp 553–558.

325. A. Wiles, "Modular Elliptic Curves and Fermat's Last Theorem", *Annals of Mathematics*, **141** (1995), pp 443–551.

326. H. S. Wilf, *Algorithms and Complexity*, 2nd Edition, A. K. Peters, 2002.

327. H. C. Williams, "A Modification of the RSA Public-Key Encryption Procedure", *IEEE Transactions on Information Theory*, **26**, (1980), pp 726–729.

328. H. C. Williams, "The Influence of Computers in the Development of Number Theory", *Computers & Mathematics with Applications*, **8**, 2(1982), pp 75–93.

329. H. C. Williams, "A $p+1$ Method of Factoring", *Mathematics of Computation*, **39**, (1982), pp 225–234.

330. H. C. Williams, "Factoring on a Computer", *Mathematical Intelligencer*, **6**, 3(1984), pp 29–36.

331. H. C. Williams, "An M^3 Public-Key Encryption Scheme", *Advances in Cryptology - CRYPTO '85*, Lecture Notes in Computer Science **218**, Springer, 1986, pp 358–368.

332. H. C. Williams, *Édouard Lucas and Primality Testing*, John Wiley & Sons, 1998.

333. C. P. Williams and S. H. Clearwater, *Explorations in Quantum Computation*, The Electronic Library of Science (TELOS), Springer-Verlag, 1998.

334. W. Windsteiger and Bruno Buchberger, "Gröbner: A Library for Computing Gröbner Bases based on SACLIB, RISC-Linz, Austra, 16 March 1995, 106 pages.

335. S. Y. Yan, *Number Theory for Computing*, 2nd Edition, Springer-Verlag, 2002.

336. S. Y. Yan, "Computing Prime Factorization and Discrete Logarithms: From Index Calculus to Xedni Calculus", *International Journal of Computer Mathematics*, **80**, 5(2003), pp 573–590.

337. S. Y. Yan, *Primality Testing and Integer Factorization in Public-Key Cryptography*, Advances in Information Security **11**, Kluwer, 2004.

338. S. M. Yen, S. Kim, S. Lim and S Moon, "RSA Speedup with Residue Number System Immune against Hardware Fault Cryptanalysis", *Information Security and Cryptology - ICISC 2001*, Lecture Notes in Computer Science **2288**, Springer, 2002, pp 1-74.

Index

About the Author

DR. SONG Y. YAN is currently Professor of Computer Science and Mathematics and Director of the Institute for Research in Applicable Computing at the University of Bedfordshire, England. He also held various visiting professorships in the Center for Discrete Mathematics and Theoretical Computer Science (DIMACS) of Rutgers/Princeton University, Columbia University, University of Toronto, Australian National University, and South China University of Technology. Most recently, he is Visiting Professor at the Massachusetts Institute of Technology. Majored in both Computer Science and Mathematics, he obtained a PhD in Number Theory in the Department of Mathematics at the University of York, England. His current research and teaching interests include Number Theory, Complexity Theory, Coding Theory, Cryptography and Information Security. He published, among others, the following three well-received books in number theory and cryptography:

[1] *Number Theory for Computing*, Springer, First Edition, 2000; Second Edition, 2002; Polish Translation, 2006 (Polish Scientific Publishers PWN); Chinese Translation, 2007 (Tsinghua University Press).

[2] *Primality Testing and Integer Factorization in Public-Key Cryptography*, Springer, First Edition, 2004; Second Edition, 2008.

[3] *Perfect, Amicable and Sociable Numbers: A Computational Approach*, World Scientific, 1996.